高等职业院校电力技术类专业系列教材

电气设备检修（一）
——基础知识类（中英双语）

主　编　姜聿涵　刘　燕
副主编　欧阳仁乐　穆　舟　赵胜霞
参　编　华　章　陈　丽　胡清灵　钟运来　廖翔志　程　铭

西南交通大学出版社
·成　都·

图书在版编目（CIP）数据

电气设备检修. 一，基础知识类：汉、英 / 姜聿涵，刘燕主编. -- 成都：西南交通大学出版社，2023.11
ISBN 978-7-5643-9582-7

Ⅰ. ①电… Ⅱ. ①姜… ②刘… Ⅲ. ①电气设备 – 设备检修 – 高等职业教育 – 教材 – 汉、英 Ⅳ. ①TM64

中国国家版本馆 CIP 数据核字（2023）第 229772 号

Dianqi Shebei Jianxiu (Yi)
—Jichu Zhishi Lei (Zhong-Ying Shuangyu)

电气设备检修（一）——基础知识类（中英双语）	主编	姜聿涵 刘燕	策划编辑	李芳芳 张少华
			责任编辑	孟媛
			封面设计	吴兵

印张：10　字数：285千	出版发行：西南交通大学出版社
成品尺寸：185 mm × 260 mm	网址：http://www.xnjdcbs.com
版次：2023年11月第1版	地址：四川省成都市二环路北一段111号西南交通大学创新大厦21楼
印次：2023年11月第1次	邮政编码：610031
印刷：四川玖艺呈现印刷有限公司	营销部电话：028-87600564　028-87600533
书号：ISBN 978-7-5643-9582-7	定价：40.00元

图书如有印装质量问题　本社负责退换
版权所有　盗版必究　举报电话：028-87600562

FOREWORD

为贯彻《国务院关于加快发展现代职业教育的决定》文件精神，更好地满足高等职业教育高质量发展的需要，实现教学内容由知识本位向能力本位的转变，结合人力资源和社会保障部国家职业技能标准《变电设备检修工》的职业能力要求，以提高电气设备检修人员的技术技能水平为目标，为满足市场和企业不断发展的岗位需求，编写了本套教材。本套教材响应"一带一路"建设号召，服务国家战略方针，深化电力行业"一带一路"交流合作。本套教材具备代表性，基础性，针对性及普遍性等特点。本教材的出版有利于拓展国际业务交流，加强电力文化输出，提升国家影响力。

本套教材共分为3本，利用"校企一体，师资互用"机制，践行"双师"结构与"双师"资质，与生产企业共同编制，由专业教师和企业教师一同进行课程内容深度分析。本套教材的编制，旨在以学生为中心，落实立德树人的根本任务，在使学生掌握相关职业资格技能鉴定或教育部"1+X"变电设备检修职业技能等级考试标准制度试点职业技能等级所需的知识和技术技能，满足相关工种的中级工或以上要求的同时，培养学生吃苦耐劳、团结协作、可迁移可转化、注重安全等职业素养和行为习惯，弘扬工匠精神，达到"德技并修、理实并重、手脑并用、讲赛并行、工学结合"的要求。

本书是电气设备检修课程的基础组成部分，根据理实一体化教学的需要，以项目导向、任务驱动为主线，学习内容遵循由浅入深、循序渐进的原则，采用教室+实训现场的理论实践相结合的教学方法，充分体现了教、学、做一体化。本书在编写过程中突出"工作任务导向、规范作业流程、理论知识够用、突出技能实训"的思想，强调安全作业和标准化作业。全书实操内容较为典型，按教学项目和教学模块设计，突出技能训练，使学员通过对技术工作的任务、过程、环境所进行的整体化感悟和反思，实现知识与技能、过程与方法、情感态度和价值观学习的统一。

本书由国网四川省电力公司技能培训中心（四川电力职业技术学院）姜聿涵、刘燕担任主编，欧阳仁乐、穆舟、赵胜霞担任副主编。全书编写分工如下：项目一电气主接线由赵胜霞、姜聿涵、钟运来编写，项目二电气设备检修标准化作业由华章、穆舟、廖翔志编写，项目三电气主接线的设计由欧阳仁乐、程铭、刘燕编写，项目四电气设备的选择由陈丽、胡清灵编写，姜聿涵负责全书内容的审定。

本书由国网四川省电力公司乐山供电公司高级技师李运涛主审，攀枝花供电公司高级技师唐启刚、遂宁供电公司高级技师赵安参与评审。在编写过程中得到国网四川省电力公司技能培训中心（四川电力职业技术学院）汤晓青副教授和电网检修部（电力设备技术系）同事的大力支持，在此表示衷心的感谢！

限于编者水平，书中不足和错误之处在所难免，恳请读者批评指正，不胜感激。

编　者

2023 年 5 月

CONTENTS

项目一 电气主接线 ········· 001
   模块一 概 述 ········· 001
   Module 1 Overview ········· 002
   模块二 单母线接线类型 ········· 003
   Module 2 Single-bus Wiring Tpye ········· 007
   模块三 双母线接线类型 ········· 013
   Module 3 Double-bus Wiring Tpye ········· 021
   模块四 无母线接线 ········· 030
   Module 4 Bus-free Wiring ········· 034
      任务一 变电站现场主接线认知 ········· 038
      Task 1 Cognition of Main Wiring in Transformer Substation ········· 040

项目二 电气设备检修标准化作业 ········· 043
   模块一 现场危险点辨识 ········· 043
   Module 1 Identification of On-site Hazards ········· 046
   模块二 现场危险点工作案例 ········· 049
   Module 2 Cases of On-site Hazards ········· 054
   模块三 标准化作业 ········· 062
   Module 3 Standardized Operation ········· 076

项目三 电气主接线的设计 ········· 101
   模块一 设计的基本要求 ········· 101
   Module 1 Basic Requirements of Design ········· 105
   模块二 设计程序和步骤 ········· 109
   Module 2 Design Procedures and Steps ········· 112
   模块三 主变压器的选择 ········· 116
   Module 3 Selection of Main Transformer ········· 119

项目四 电气设备的选择 ··· 124
    模块一 短路电流的效应 ··· 124
    Module 1　Effect of Short-circuit Current ································· 127
    模块二 电气设备选择的一般条件 ··· 130
    Module 2　General Conditions for Selection of Electrical Equipment ·········· 132
    模块三 高压电气设备的选择 ·· 135
    Module 3　Selection of HV Electrical Equipment ····························· 142

参考文献 ·· 153

项目一　电气主接线

模块一　概　述

一、电气主接线基本概念

电气主接线又称为电气一次接线，它是将电气设备以规定的图形和文字符号，按电能生产、传输、分配顺序及相关要求绘制的单相接线图。主接线代表了发电厂或变电站高电压、大电流的电气部分主体结构，是电力系统网络结构的重要组成部分。它直接影响电力生产运行的可靠性、灵活性，同时对电气设备选择、配电装置布置、继电保护、自动装置和控制方式等诸多方面都有决定性的关系。电气主接线有三大基本要求：可靠性、灵活性、经济性。安全可靠是电力生产的首要任务，保证供电可靠是电气主接线最基本的要求。电气主接线应能适应各种运行状态，并能灵活地进行运行方式转换。经济性往往与可靠性相矛盾，因此主接线要在满足可靠性和灵活性的前提下做到经济合理。

二、主接线的基本形式

主接线的基本接线形式就是主要电气设备常用的几种连接方式，以电源和出线为主体。由于各个发电厂或变电站的出线回路数和电源数不同，且每路馈线所传输的功率也不一样，因而为便于电能的汇集和分配，在进出线数较多时（一般超过 4 回路），采用母线作为中间环节，可使接线简单清晰，运行方便，有利于安装和扩建。而与有母线的接线相比，无汇流母线的接线使用开关电器较少，配电装置占地面积较小，通常用于进出线回路少，不再扩建和发展的发电厂或变电站。

电气主接线可分为有母线、无母线两类，其中有母线的接线形式包括单母线接线、双母线接线；无母线接线形式包括桥形接线、多角形接线和单元接线。电气主接线的基本形式如图 1-1 所示。

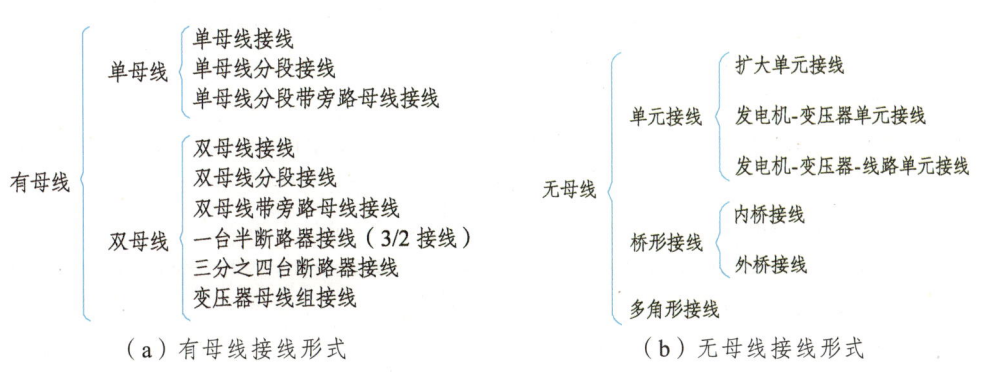

图 1-1　电气主接线分类

Program 1 Main Electrical Wiring

Module 1 Overview

1. Basic Concept

The main electrical wiring, also called the primary electrical wiring, is a single-phase wiring diagram of electrical equipment drawn according to the production, transmission, distribution sequence and related requirements with specified graphics and text symbols. The main wiring represents the main structure of the electrical part of the power plant or substation with high voltage and current, serving as an important part of the network structure of the power system. It directly affects the reliability and flexibility of power production and operation, and has a decisive relationship with many aspects such as the selection of electrical equipment, Power Distribution Unit (PDU) layout, relay protection, automatic device, and control mode.

The main electrical wiring has three basic requirements: reliability, flexibility, and economical efficiency. Safety and reliability is the primary task of power production, while the most basic requirement for main electrical wiring is to guarantee the reliable power supply. The main electrical wiring shall be able to adapt to different operating states, and can flexibly change the operating mode. Economy is often at odds with reliability, so the main wiring should be economical and reasonable under the premise of reliability and flexibility.

2. Basic Forms of Main Wiring

The basic forms of main wiring are the common connection methods for the main electrical equipment, with the power supply and the outgoing line as the main body. Each power plant or substation has different numbers of outgoing circuits and power supplies, and the power transmitted by each feeder is not the same. In case of many incoming and outgoing lines (generally more than 4 circuits), the bus is used as the intermediate link to facilitate the collection and distribution of electric energy. This can make the wiring simple and clear, operation-friendly, and conducive to installation and expansion. Compared with the bus wiring, the wiring without bus rods boasts fewer switching devices and a smaller PDU footprint area. It is usually used for power plants or substations with fewer incoming and outgoing lines and no further expansion and development.

The main electrical wiring can be divided into two types: bus wiring and non-bus wiring. The bus wiring forms include single bus and double bus wiring. The non-bus wiring forms include bridge wiring, polygonal wiring and unit wiring. Fig. 1-1 shows the specific categories.

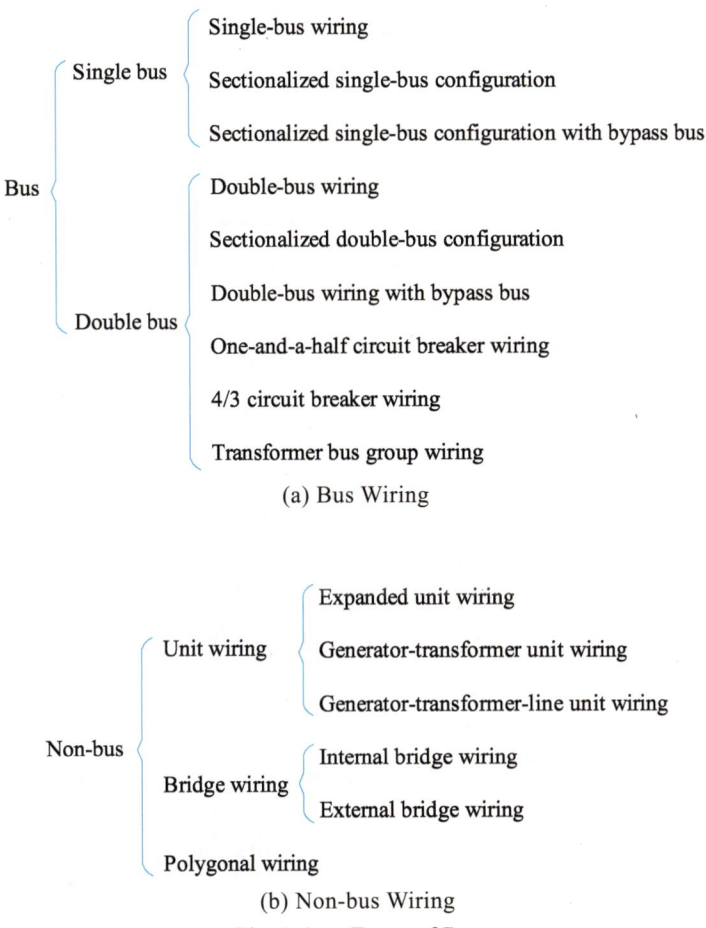

Fig. 1-1　Types of Bus

模块二　单母线接线类型

一、单母线接线

（一）接线特点

单母线接线如图 1-2 所示。单母线供电电源在发电厂是发电机或变压器，在变电站是变压器或高压进线回路。母线既可保证电源并列工作，又能使任一条出线都可以从任一个电源获得电能。各出线回路输送功率不一定相等，应尽可能使负荷均衡地分配于母线上，以减少功率在母线上的传输。

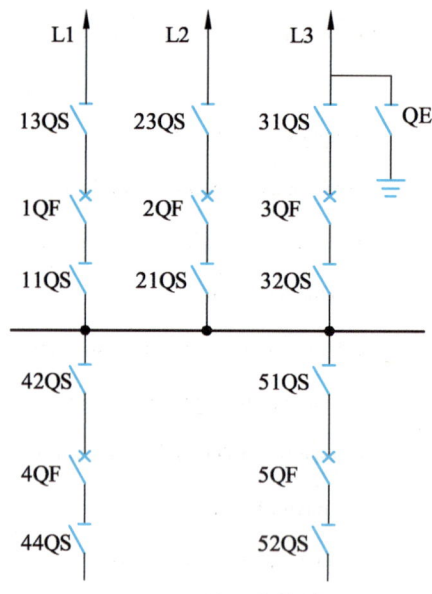

图 1-2 单母线接线

单母线接线的特点是每一回路均有断路器 QF 和隔离开关 QS。断路器用于在正常或故障情况下接通与断开电路。断路器两侧装有隔离开关,用于停电检修断路器时作为明显断开点以隔离电压。在电源回路中,若断路器断开之后,电源不可能向外输送电能时,断路器与电源之间可以不装隔离开关,如发电机出口。若线路对侧无电源,则线路侧可不装设隔离开关。

高压隔离开关一般有主闸刀和接地开关。QE 是线路隔离开关的接地开关,用于线路检修时替代临时安全接地线。为避免发生接地开关接地状态下误合主闸刀的事故,主闸刀与接地开关之间设有机械联锁装置。优点:接线简单,操作方便,设备少、经济性好,并且母线便于向两端延伸,扩建方便。

缺点:(1)可靠性差。母线或母线隔离开关检修或者故障时,所有回路都要停止运行,造成全厂(站)长期停电;(2)调度不方便。电源只能并列运行,并且线路侧发生短路时,有较大的短路电流。

综上,这种接线形式一般只用在出线回路少,并且没有重要负荷的发电厂和变电站中。

(二)典型操作

1. 线路停电操作

以 L1 线路停电为例,操作步骤是:断开 1QF 断路器,检查 1QF 是否确实断开,断开 13QS 隔离开关,断开 11QS 隔离开关。

2. 线路送电操作

以 L1 线路送电为例,操作步骤是:检查 1QF 是否确实断开,合上 11QS 隔离开关,合上 13QS 隔离开关,合上 1QF 断路器。这样的操作遵循了两条原则:一是防止带负荷拉合隔离开关；二是防止在断路器出于合闸状态下，误操作隔离开关的事故不发生在母线隔离开关上，以避免误操作的电弧引起母线短路事故。

二、单母线分段接线

单母线分段接线如图 1-3 所示。正常运行时，单母线分段接线有如下两种运行方式。

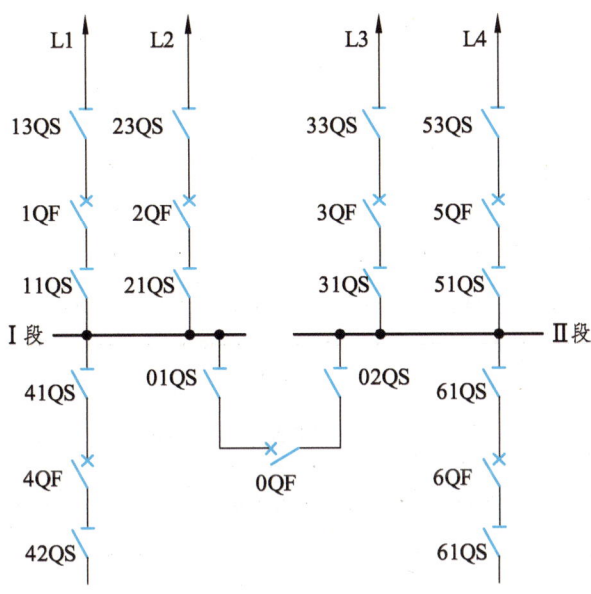

图 1-3　单母线分段

1. 分段断路器闭合运行

正常运行时分段断路器 0QF 闭合，两个电源分别接在两段母线上。两段母线上的负荷应均匀分配，以使两段母线上的电压均衡。在运行中，当任一段母线发生故障时，继电保护装置动作跳开分段断路器和接至该母线段上的电源断路器，另一段则继续供电。有一个电源故障时，仍可以使两段母线都有电，可靠性比较好。但是线路故障时短路电流较大。

2. 分段断路器断开运行

正常运行时分段断路器 0QF 断开，两段母线上的电压可不相同。每个电源只向接至本段母线上的引出线供电。当任一电源出现故障，接该电源的母线停电，导致部分用户停电。为了解决这个问题，可以在 0QF 处装设备自投装置，或者重要用户可以从两段母线引接，采用双回路供电。分段断路器断开运行的优点是可以限制短路电流。

该接线适用于：小容量发电厂的发电机电压配电装置，一般每段母线上所接发电容量为 12 MW 左右，每段母线上出线不多于 5 回路；变电站有两台主变压器时的 6～10 kV 配电装置；35～63 kV 配电装置出线 4～8 回路；110～220 kV 配电装置出线 3～4 回路。

三、单母线分段带旁路母线接线

断路器经过长期运行和切断数次短路电流后都需要检修。为了使采用单母线分段或双母线的配电装置检修断路器时，不致中断该回路供电，可增设旁路母线。

通常，旁路母线有三种接线方式：有专用旁路断路器的旁路母线接线，母联断路器兼作旁路断路器的旁路母线接线，用分段断路器兼作旁路断路器的旁路母线接线。

（一）接线特点

图 1-4 为单母线分段带旁路接线的一种情况。旁路母线经旁路断路器接至Ⅰ、Ⅱ段母线上。正常运行时，90QF 回路以及旁路母线处于冷备用状态。

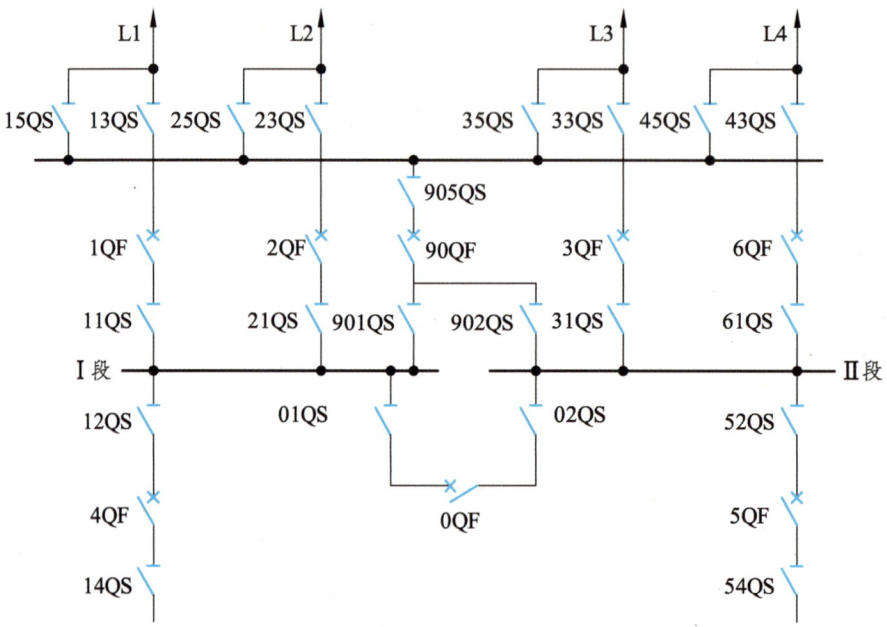

图 1-4 单母线分段带旁路母线接线

当出线回路数不多时，旁路断路器利用率不高，可与分段断路器合用，有以下两种形式：

1. 分段断路器兼作旁路断路器

如图 1-5 所示，从分段断路器 0QF 的隔离开关内侧引接联络隔离开关 05QS 和 06QS 至旁路母线，在分段工作母线之间再加两组串联的分段隔离开关 03QS 和 04QS。正常运行时，分段断路器 0QF 及其两侧隔离开关 03QS 和 04QS 处于接通位置，联络隔离开关 05QS 和 06QS 处于断开位置，分段隔离开关 01QS 和 02QS 中，一组断开，一组闭合，旁路母线不带电。

2. 旁路断路器兼作分段断路器

如图 1-6 所示。正常运行时，两分段隔离开关 01QS、02QS 一个投入一个断开，两段母线通过 901QS、90QF、905QS、旁路母线、03QS 相连接，90QF 起分段断路器作用。

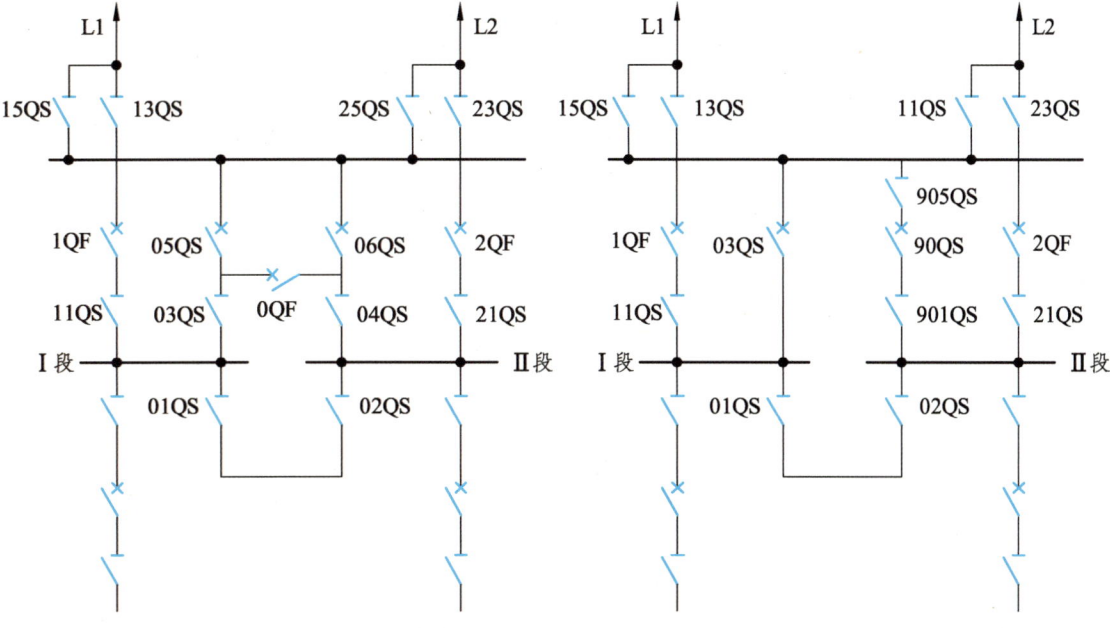

图 1-5 分段断路兼作旁路断路器　　　　图 1-6 旁路断路器兼作分段断路器

单母线分段带有专用旁路断路器的旁路母线极大地提高了可靠性，但是增加了一台旁路断路器的投资。分段断路器兼作旁路断路器的接线，可以减少设备、节省投资。

（二）典型操作

以图 1-4 为例，检修线路 L1 的断路器 1QF 时，要求线路不停电，其操作顺序如下：检查 90QF 确已断开，合上 901QS，合上 905QS，合上 90QF；检查旁路母线电压正常，断开 90QF，合上 15QS，合上 90QF；检查 90QF 三相电流平衡，断开 1QF，断开 13QS，断开 11QS；然后按检修要求做好安全措施，即可对 1QF 进行检修，而整个过程 L1 线路不停电。

Module 2　Single-bus Wiring Tpye

1. Single-bus Wiring

1) Wiring Characteristics

Fig. 1-2 shows the single-bus wiring.

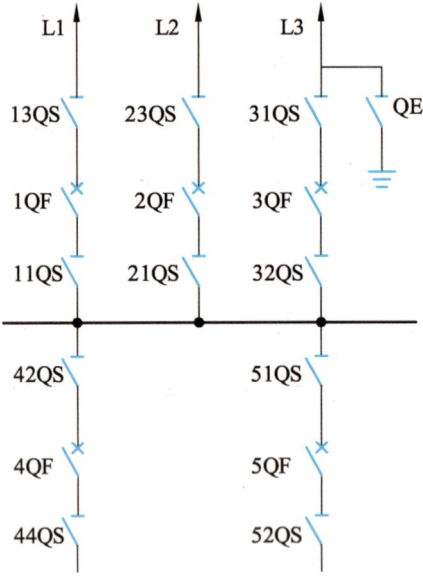

Fig. 1-2　Single-bus Wiring

　　The single-bus power supply is a generator or transformer in a power plant, and a transformer or high voltage incoming circuit in a substation. The bus can ensure that the power supply works side by side, and any outgoing line can get power from either source. The transmission power of each outgoing circuit is not necessarily equal, and the load should be evenly distributed on the bus as far as possible to reduce the transmission of power on the bus.

　　The single-bus wiring is characterized by the circuit breaker QF and disconnector QS on each circuit. The circuit breaker is used to make and break the circuit in normal or fault conditions, both sides of the circuit breaker are equipped with disconnectors to be used as an obvious break point to isolate voltage during interruption maintenance of circuit breakers. The disconnector near the bus side is called the bus side disconnector (such as 11QS), and the disconnector near the outgoing line side is called the line side disconnector (such as 13QS). The first few digits of the disconnector number in the main wiring device number are the same as the branch circuit breaker number. The last digit of the line side disconnector number is 3, and the last digit of the bus side disconnector number is 1 (1 and 2 in case of double bus). In the power supply circuit, if the circuit breaker is turned off, the power supply cannot supply power outward, disconnector may not be installed between the circuit breaker and the power supply, such as the generator outlet. If there is no power supply on the opposite side of the line, disconnector may not be installed on the line side.

　　The high-voltage disconnector generally have a main knife switch and a grounding switch. QE is the grounding switch of the line disconnector, which is used to replace the temporary safety grounding wire during line maintenance. In order to avoid the accident of misclosing the main knife switch when the grounding switch is grounded, a mechanical interlocking device is arranged between the main knife switch and the grounding switch.

Advantages: simple wiring, convenient operation, few equipment, good economy, and the bus is easy to extend to both ends, easy expansion.

Disadvantages:

(1) Poor reliability. When the bus or bus disconnector is repaired or faulty, all circuits has to stop running, resulting in long-term outage of the whole power plant (station).

(2) Inconvenient scheduling. The power supply can only run side by side. In case of a short circuit on the line side, the short-circuit current is long.

In summary, this type of wiring is generally used in power plants and substations with few outgoing circuits and no important loads.

2) Typical Operation

(1) Line outage operation.

Taking line L1 outage as an example, the operation steps are as follows: Disconnect the circuit breaker 1QF, check that the 1QF is disconnected indeed, disconnect the disconnector 13QS, and disconnect the disconnector 11QS.

(2) Line energization operation.

Taking line L1 energization as an example, the operation steps are as follows: Check that the 1QF is disconnected indeed, close the disconnector 11QS, close the disconnector 13QS, and close the circuit breaker 1QF.

This operation follows two principles: First, to prevent the switching of disconnector with load; Second, to prevent the accident of misoperation of the disconnector on the bus disconnector when the circuit breaker is in the closed status, so as to avoid the bus short circuit caused by the misoperation of arc.

2. Sectionalized Single-bus Configuration

The sectionalized single-bus configuration is as shown in Fig. 1-3. In normal operation, there are two modes of operation for sectionalized single-bus configuration:

(1) Closed operation of section circuit breaker.

During normal operation, the section circuit breaker 0QF is closed, and the two power supplies are connected to two sections of the bus respectively. The loads on two sections of the bus shall be evenly distributed so that their voltage is equalized. In operation, when any section of the bus fails, the relay protection device will operates to disconnect the section circuit breaker and the power circuit breaker connected to this section of the bus, and the other section keeps the power on. In case of a power failure, both sections of bus can still be powered, proving good reliability. However, the short circuit current is high when the line fails.

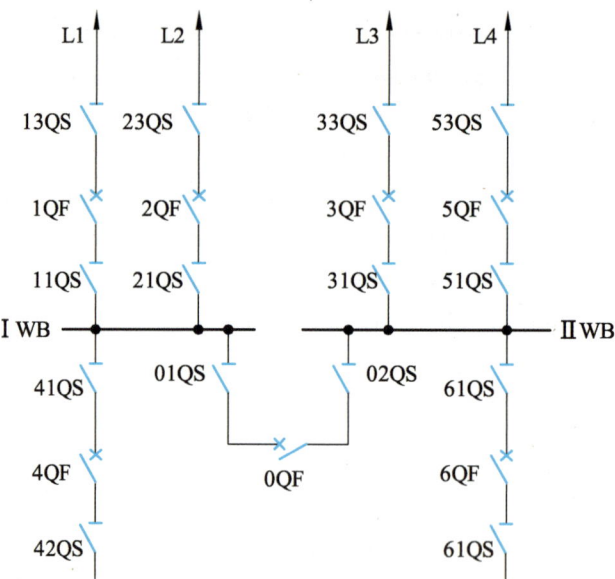

Fig. 1-3　Sectionalized Single-bus Configuration

(2) Opening operation of section circuit breaker.

During normal operation, the section circuit breaker 0QF is disconnected, and the voltages on the two sections of bus are different. Each power supply powers only to the outgoing line connected to this section of bus. When any power supply fails, the bus connecting to the power supply is cut off, leading to outage of some users. To solve this problem, an auto-switchover unit can be installed at 0QF, or important users may benefit from the double circuit power supply from two sections of bus. The opening operation of section circuit breaker can limit the short-circuit current.

The wiring is applicable to generator PDUs for small-capacity power plants, generally each section of the bus is connected to the power generation capacity of about 12MW, each section of the bus line is not more than 5 circuits are led out of the bus; 6~10 kV power distribution unit when the substation has two main transformers; 4~8 circuits out from the 35~63 kV power distribution unit; and 3~4 circuits out from the 110~220 kV power distribution unit.

3. Sectionalized Single-bus Configuration with Bypass Bus

The circuit breaker needs to be overhauled after long-term operation and cut-off times of short circuit current. The bypass bus can be provided to ensure that the power supply of this circuit will not be interrupted when the circuit breaker is repaired by the power distribution unit with sectionalized single bus or double bus.

Generally, the bypass bus is connected in three ways: the bypass bus wiring with a dedicated bypass circuit breaker, the bypass bus wiring with bus-tie circuit breaker doubling as the bypass circuit breaker, and the bypass bus wiring with the section circuit breaker doubling as the bypass circuit breaker.

1) Configuration Characteristics

Fig. 1-4 shows a scenario of sectionalized single-bus configuration with bypass bus. The bypass bus is connected to the Section I and Section II of the bus through the bypass circuit breaker. During normal operation, the circuit 90QF and the bypass bus are in the cold standby state.

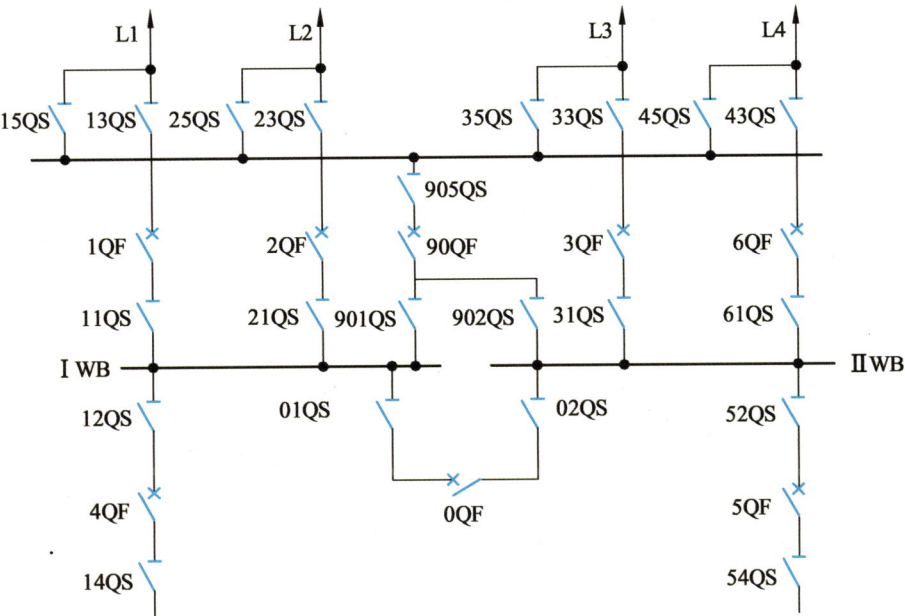

Fig. 1-4　Sectionalized Single-bus Configuration with Bypass Bus

If the outgoing circuits is few, the bypass circuit breaker is less used and can be shared with the section circuit breaker, there are two forms:

(1) The section circuit breaker doubling as a bypass circuit breaker.

As shown in Fig. 1-5, connect the disconnectors 05QS and 06QS from the inner side of the disconnector of the section circuit breaker 0QF to the bypass bus, and add two groups of section disconnectors 03QS and 04QS in series between the sectionalized operating buses. During normal operation, the section circuit breaker 0QF and its disconnectors 03QS and 04QS at both sides are in the on-position, and the interconnection isolators 05QS and 06QS are in the off-position. For the sectionalized disconnector 01QS and 02QS, one set is disconnected, and the other set is closed. The bypass bus is electrically neutral.

(2) The bypass circuit breaker is also used as the section circuit breaker.

As shown in Fig. 1-6, during normal operation, there are two sectionalized disconnectors 01QS and 02QS, one of which is put into operation with the other one disconnected. Two sections of the bus are connected by 901QS, 90QF, 905QS, bypass bus and 03QS, and the 90QF serves as the section circuit breaker.

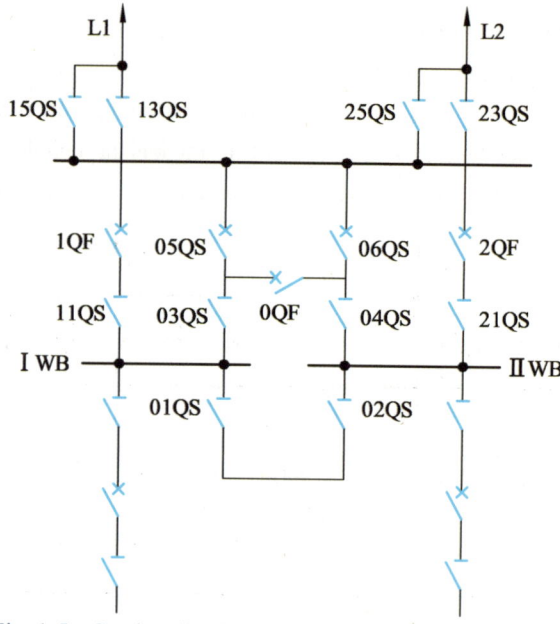

Fig. 1-5 Sectionalized single-bus wiring with bypass bus

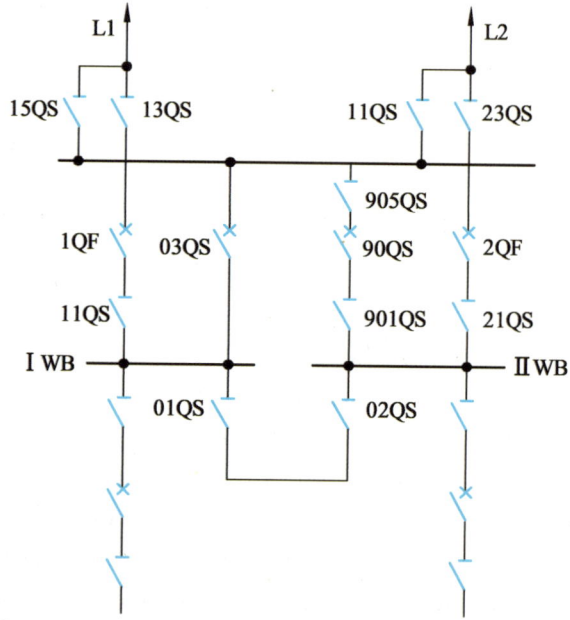

Fig. 1-6 Sectionalized Single-bus Wiring with Bypass Bus

Single-bus sectionalized bypass bus with dedicated bypass circuit breaker greatly improves the reliability, but increases the investment in a bypass circuit breaker. The section circuit breaker also serves as the wiring of the bypass circuit breaker, which can reduce equipment and save investment.

2) Typical Operation

Fig. 1-4 is taken as an example. During maintenance of the circuit breaker 1QF of line

L1, it is required that the line is not power-off. The operation sequence is as follows:

Check that the 90QF is disconnected indeed, switch on the 901QS, 905QS and 90QF. Check that the voltage of the bypass bus is normal, disconnect the 90QF, switch on the 15QS and 90QF. Check that the 90QF has balanced three-phase current, disconnect the 1QF, disconnect the 13QS and 11QS, and then maintain the 1QF with safety measures taken as per maintenance requirements. In the whole process, the line L1 is not power-off.

模块三　双母线接线类型

一、双母线接线

（一）接线特点

不分段的双母线接线如图 1-7 所示。这种接线有两组母线（Ⅰ段和Ⅱ段），在两组母线之间通过母线联络断路器 0QF（以下简称母联断路器）连接；每一条引出线（L1、L2、L3、L4）和电源支路（5QF、6QF）都经一台断路器与两组母线隔离开关分别接至两组母线上。

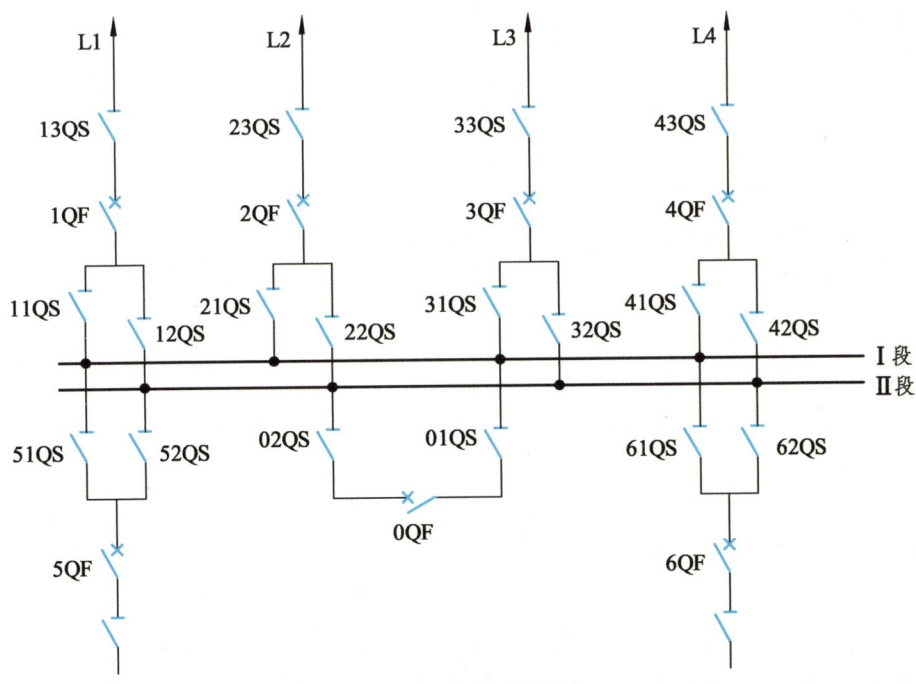

图 1-7　双母线接线

双母线接线的主要优点是供电可靠、运行灵活、检修方便、易于扩建，在大、中型发电厂和枢纽变电站中广为采用。

（二）典型操作

以下操作均以图 1-7 为例。

1. Ⅰ母线运行转检修操作

（1）正常运行方式：两组母线并联运行，L1、L3、5QF 接Ⅰ母线，L2、L4、6QF 接Ⅱ母线。

操作步骤：

确认 0QF 在合闸运行，取下 0QF 操作电源保险，合上 52QS，断开 51QS，合上 12QS，断开 11QS，合上 32QS，断开 31QS，合上 0QF 操作电源保险。然后断开 0QF，检查 0QF 确已断开，断开 01QS，断开 02QS。接着退出Ⅰ母线电压互感器，按检修要求做好安全措施，即可对Ⅰ母线进行检修，而整个操作过程没有任何回路停电。

在此过程中，操作刀闸之前取下 0QF 操作电源保险，是为了在操作过程中母联断路器 0QF 不跳闸，确保所操作刀闸两侧可靠等电位。因为如果在操作过程中母联断路器跳闸，可能会导致带负荷断开（合上）隔离开关，造成事故。

（2）正常运行方式：Ⅰ母线为工作母线，Ⅱ母线为备用母线。

操作步骤：

依次合上母联隔离开关 01QS 和 02QS，再合上母联断路器 0QF，用母联断路器向备用母线充电，检验备用母线是否完好：若备用母线存在短路故障，母联断路器立即跳闸；若备用母线完好时，合上母联断路器后不跳闸。

然后取下 0QF 操作电源保险，合上 52QS，断开 51QS，合上 62QS，断开 61QS，合上 12QS，断开 11QS，合上 22QS，断开 21QS，合上 32QS，断开 31QS，合上 42QS，断开 41QS，投上 0QF 操作电源保险，由于母联断路器连接两套母线，所以依次合上、断开以上隔离开关只是转移电流，而不会产生电弧。

断开母联断路器 0QF，依次断开母联隔离开关 01QS 和 02QS。至此，Ⅱ母线转换为工作母线，Ⅰ母线转换为备用母线。在上述操作过程中，任一回路的工作均未受到影响。

2. 51QS 隔离开关检修

正常运行方式：两组母线并联运行，L1、L3、5QF 接Ⅰ母线，L2、L4、6QF 接Ⅱ母线。

操作步骤：

只需将 L1、L3 线路倒换到Ⅱ母线上运行，然后断开该回路和与此隔离开关相连接的Ⅰ母线，并做好安全措施，该隔离开关就可以停电检修。具体操作步骤参考操作上文"Ⅰ母线运行转检修操作"。

3. L1 线路断路器 1QF 拒动，利用母联断路器切断 L1 线路

正常运行方式是：两组母线并联运行，L1、L3、5QF 接Ⅰ母线，L2、L4、6QF 接Ⅱ母线。

操作步骤：

首先利用倒母线的方式，将 L3 回路和 5QF 回路从Ⅰ母线上倒到Ⅱ母线上运行，这时 L1 线路、1QF、Ⅰ母线、母联、Ⅱ母线形成串联供电电路。然后断开母联断路器 0QF 切断电路，即可保证线路 L1 可靠切断。具体操作步骤可以参考前文相关操作自己练习。

二、双母线分段接线

双母线分段接线如图 1-8 所示，Ⅰ母线用分段断路器 00QF 分为两段，每段母线与Ⅱ母线之间分别通过母联断路器 01QF、02QF 连接。这种接线较双母线接线具有更高的可靠性和更大的灵活性。当Ⅰ组母线工作、Ⅱ组母线备用时，它具有单母线分段接线的特点。Ⅰ组母线的任一分段检修时，将该段母线所连接的支路倒至备用母线上运行，仍能保持单母线分段运行的特点。当具有三个或三个以上电源时，可将电源分别接到Ⅰ组的两段母线和Ⅱ组母线上，用母联断路器连通Ⅱ组母线与Ⅰ组某一个分段母线，构成单母线分三段运行，可进一步提高供电可靠性。

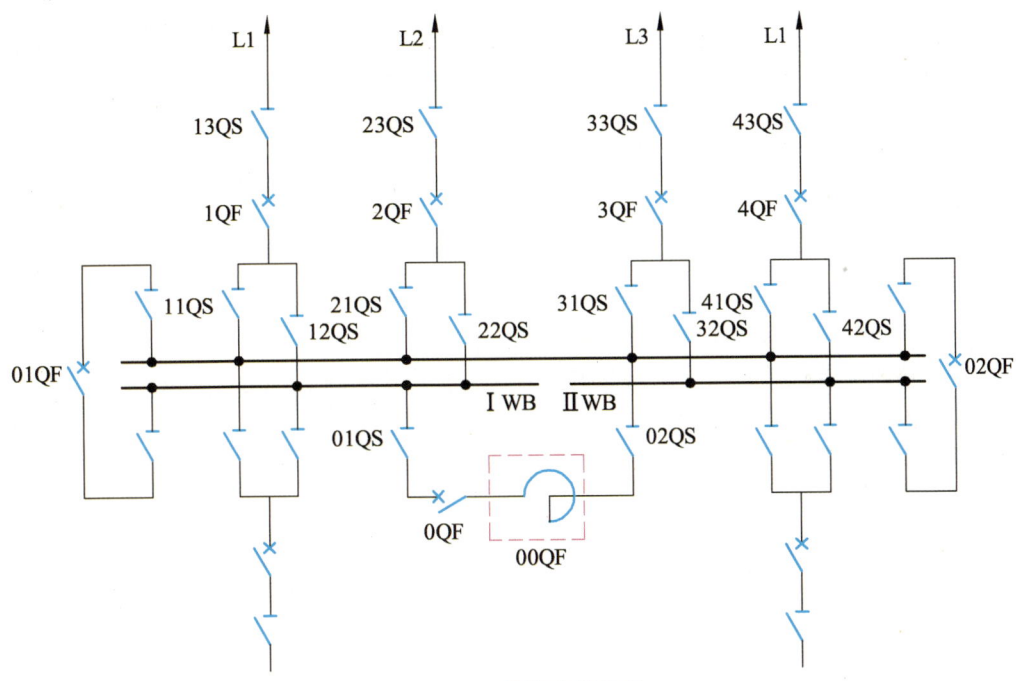

图 1-8 双母线分段接线

双母线分段接线比双母线接线增加了两台断路器，投资有所增加。但双母线分段不仅具有双母线的各种优点，并且任何时候都有备用母线，有较高的可靠性和灵活性。

三、双母线带旁路母线接线

（一）接线特点

有专用旁路断路器的双母线带旁路接线如图 1-9 所示。旁路断路器可代替出线断路器工作，使出线断路器检修时，线路供电不受影响。双母线带旁路接线，正常运行多采用两组母线固定连接方式，即双母线同时运行的方式，此时母联断路器处于合闸位置，并要求某些出线和电源固定连接于Ⅰ母线上，其余出线和电源连至Ⅱ母线。两组母线固定连接回路的确定既要考虑供电可靠性，又要考虑负荷的平衡，尽量使母联断路器通过的电流很小。

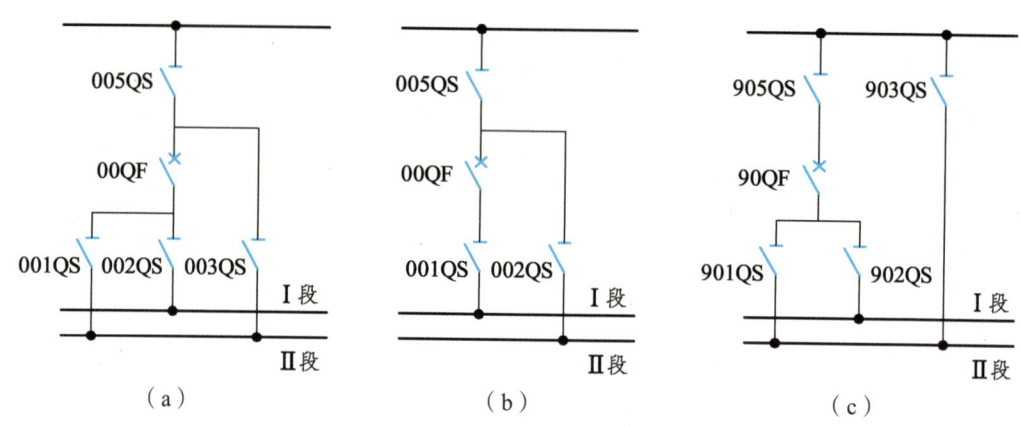

图1-9 有专用旁路断路器的双母线带旁路接线

双母线带旁路接线采用固定连接方式运行时，通常设有专用的母线差动保护装置。运行中，如果一组母线发生短路故障，则母线保护装置动作跳开与该母线连接的出线、电源和母联断路器，维持未故障母线的正常运行。然后，可按操作规程的规定将与故障母线连接的出线和电源回路倒换到未故障母线上恢复送电。

用旁路断路器代替某出线断路器供电时，应将旁路断路器90QF与该出线对应的母线隔离开关合上，以维持原有的固定连接方式。

当出线数目不多、安装专用的旁路断路器利用率不高时，为了节省资金，可采用母联断路器兼作旁路断路器的接线，具体连接如图1-9（a）、（b）、（c）所示。

图1-9（a）所示接线，按固定连接方式运行时002QS、003QS、00QF闭合，001QS、005QS断开，旁路母线不带电，旁路断路器00QF作为母联断路器运行；如果需要用00QF代替出线断路器供电时，需先将双母线的运行方式改为单母线运行，再按操作规程完成用00QF代替出线断路器的操作。

图1-9（b）所示接线，按固定连接方式运行时，001QS、00QF、002QS闭合，005QS断开，旁路母线不带电运行。用00QF代替出线断路器供电时，需先将Ⅱ母线倒换为备用母线，Ⅰ母线为工作母线，然后完成用00QF代替出线断路器的操作。

图1-9（c）所示接线，按固定连接方式运行时，902QS、90QF、905QS、903QS闭合，901QS断开，旁路母线带电运行。用90QF代替出线断路器供电时，需先将双母线的运行方式改为单母线运行，再按操作规程完成用90QF代替出线断路器的操作。

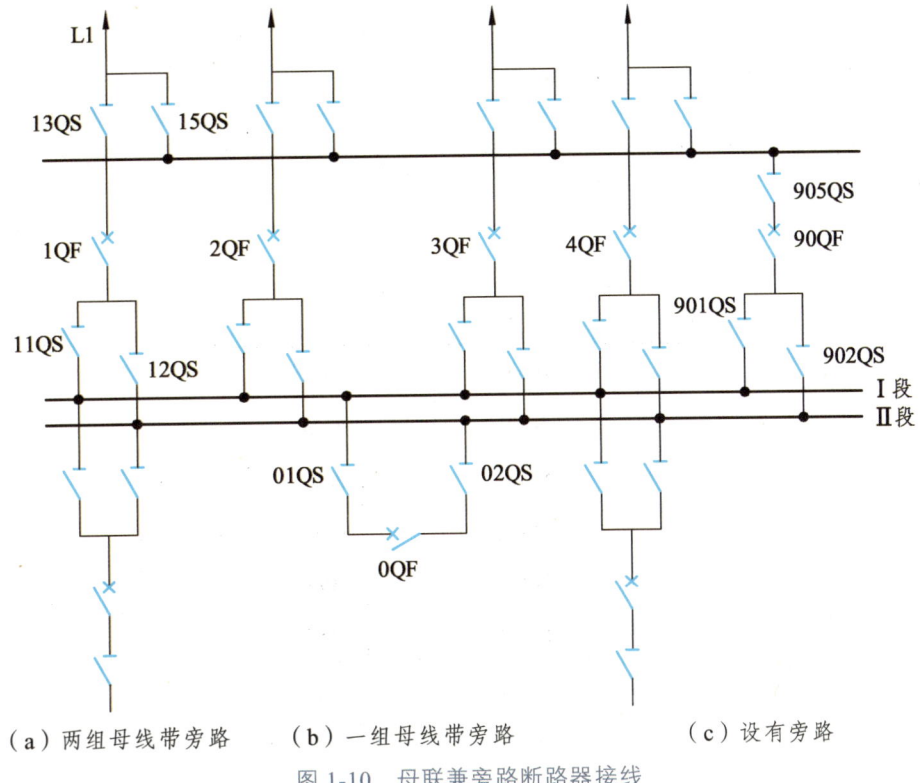

（a）两组母线带旁路　　（b）一组母线带旁路　　（c）设有旁路

图 1-10　母联兼旁路断路器接线

（二）旁路母线设置原则

110 kV 及以上的高压配电装置中，因为电压等级高、输送功率较大、送电距离较远，停电影响较大，同时高压断路器每台检修通常都需要 5~7 天的较长时间，因而允许因检修断路器而长期停电，均需设置旁母母线，检修与它相连的任一回的断路器时，该回路便可以不停电，提高了供电的可靠性。当 110 kV 出线在 6 回路及以上、220 kV 出线在 4 回路及以上时，宜采用带专用旁路断路器的旁路母线。

随着高压配电装置广泛采用的 SF_6 断路器及国产断路器质量的提高，同时系统备用容量的增加、电网结构趋于合理与联系紧密、保护双重化的完善以及设备检修逐步由计划检修向状态检修过渡，为简化接线，总的趋势是将逐步取消旁路设施。

四、一台半断路器接线

（一）接线特点

一台半断路器接线如图 1-11 所示。该接线有两组母线，每一回路经一台断路器接至一组母线，两个回路间有一台断路器联络，形成一串，每回进出线都与两台断路器相连，而同一串的两条进出线共用三台断路器，故而得名一台半断路器接线或叫作 3/2 接线。

正常运行时，两组母线同时工作，所有断路器均闭合，称为完整串运行，形成多环路状供电，具有很高的可靠性。其主要特点是：任一母线故障或检修，均不致停电；任一断路器检修也不致停电；甚至两组母线同时故障的极端情况下，功率仍能继续传输。一串中任何一台断路器退出或检修时，这种方式称为不完整串运行，此时仍不影响任何一个元件的运行。这种接线运行方便、操作简单，隔离开关只在检修时作为隔离带电设备使用。

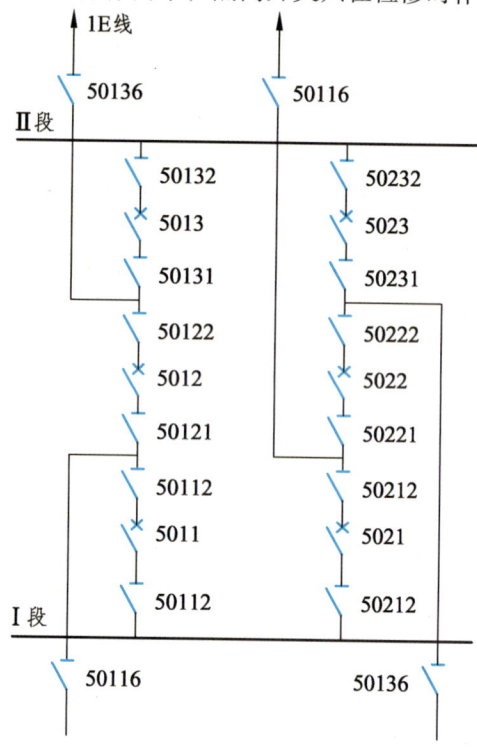

图 1-11 一台半断路器接线

（二）典型操作

1. I 母线由运行转检修

（1）断开 5011 断路器，检查 5011 断路器在分闸位。
（2）断开 5021 断路器，检查 5021 断路器在分闸位。
（3）断开 50111 隔离开关，检查 50111 隔离开关分闸到位。
（4）断开 50211 隔离开关，检查 50211 隔离开关分闸到位。
（5）进行保护的投退和安全措施后，即可对 I 母线进行检修。

2. 1E 线路由检修转运行

（1）撤出安全措施和进行保护的投退。
（2）检查 5012 断路器确实断开。
（3）检查 5013 断路器确实断开。
（4）合上 50136 隔离开关，检查 50136 隔离开关合闸到位。

(5)合上 5013 断路器,检查 5013 断路器在合闸位。
(6)合上 5012 断路器,检查 5012 断路器在合闸位。

3. 5012 断路器由运行转检修

(1)检查 5012 断路器确实断开。
(2)断开 50122 隔离开关,检查 50122 隔离开关分闸到位。
(3)断开 50121 隔离开关,检查 50121 隔离开关分闸到位。
(4)在进行保护的投退和安全措施后,即可对 5012 断路器进行检修。

(三)适用范围

一台半断路器接线,目前在国内、外已较广泛应用于大型发电厂和变电站的 330~500 kV 的配电装置中。当进出线回路数为 6 回路及以上,在系统中占重要地位时,宜采用一个半断路器接线。

五、三分之四台断路器接线

由于高压断路器造价高,为了进一步减少设备投资,把 3 条回路的进出线通过 4 台断路器接到两组母线上,构成三分之四断路器接线方式,如图 1-12 所示。

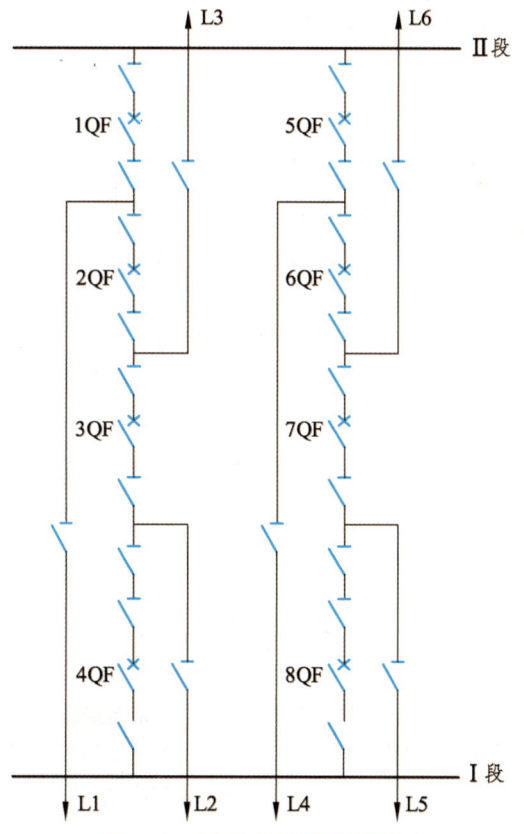

图 1-12　三分之四断路器接线

这种接线方式通常用于发电机台数（进线）大于线路（出线）数的大型水电厂，以便实现在一个串的3个回路中电源与负荷容量相互匹配。

实际运用中，可以根据电源和负荷的数量及扩建要求，采用三分之四台、一台半及两台断路器的多重连接的组合接线，有利于提高配电装置的可靠性和灵活性。

六、变压器母线组接线

除了以上常见的几种接线之外，还可以采用如图1-13所示的变压器母线组接线。这种接线变压器直接接入母线，各出线回路采用双断路器接线，如图3-13（a）所示，或者采用一个半断路器接线，如图1-13（b）所示。该种接线方式调度灵活，电源与负荷可以自由调配，安全可靠，利于扩建。

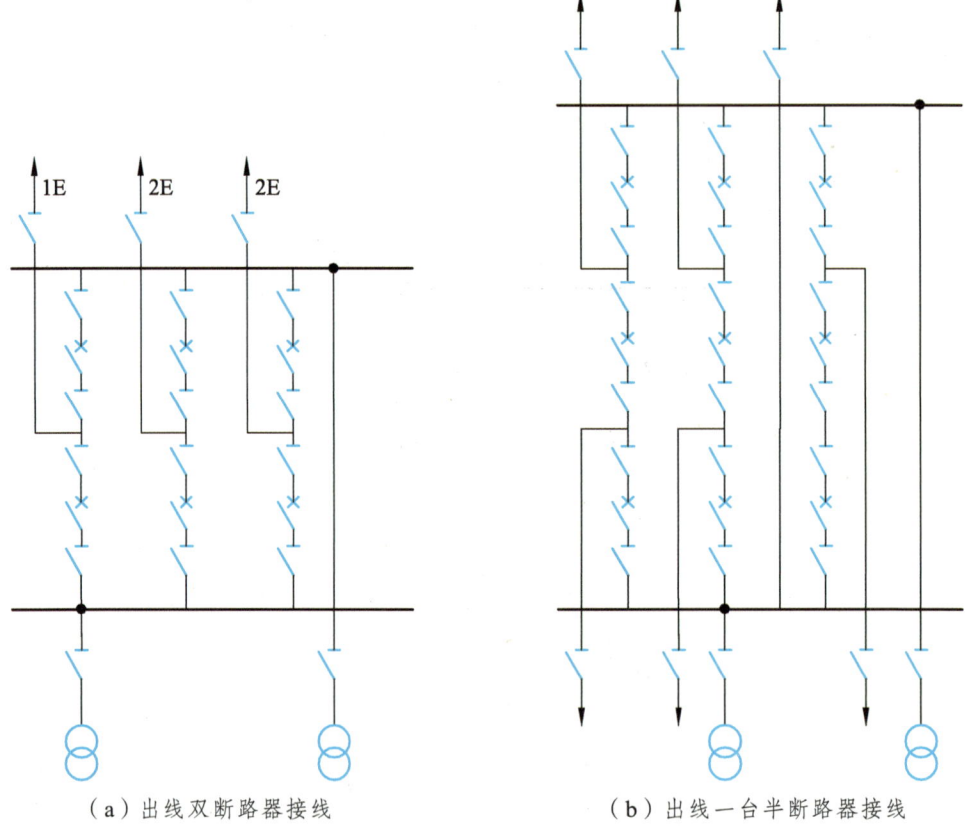

(a) 出线双断路器接线　　　　(b) 出线一台半断路器接线

图1-13　变压器母线组接线

由于变压器运行可靠性比较高，所以直接接入母线，对母线运行不产生明显的影响。一旦变压器故障，连接于母线上的断路器跳开，但不影响其他回路供电，再用隔离开关把故障变压器退出后，即可进行倒闸操作，使该母线恢复运行。这种接线适用于长距离、大容量输电线路、系统稳定性问题突出和要求线路有高度可靠性的并要求主变压器的质量可靠、故障率甚低的变电站中。

Module 3　Double-bus Wiring Tpye

1. Double-bus Wiring

1) Wiring Characteristics

Non-sectionalized double-bus wring is shown in Fig. 1-7. Such wiring has two sets of buses (IWB and IIWB) which are connected by bus-tie circuit breaker 0QF (BTCB). Each lead wire (L1, L2, L3, L4) and power supply branch (5QF, 6QF) are connected to two sets of buses respectively through a circuit breaker and two sets of bus disconnectors.

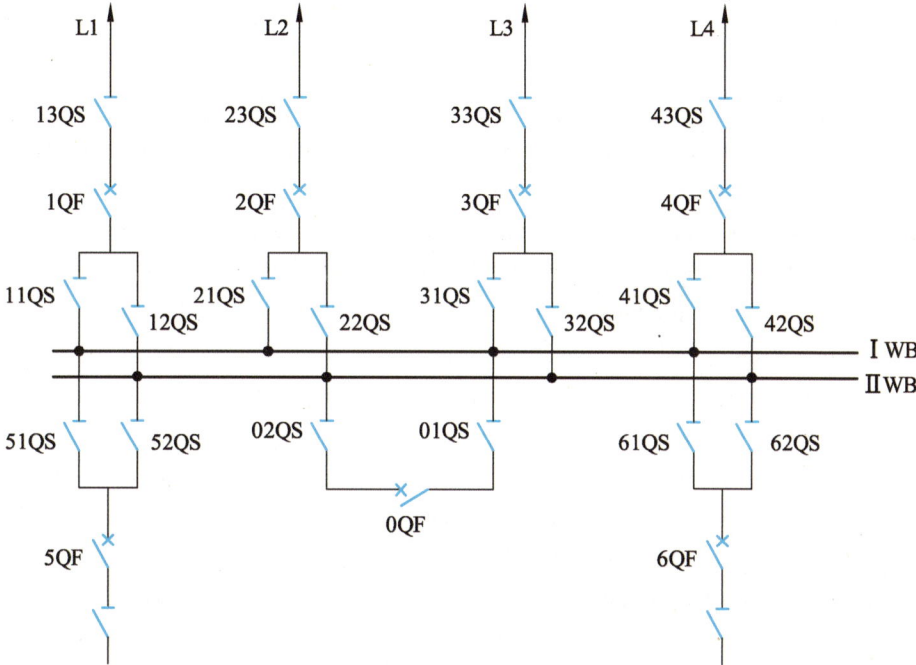

Fig. 1-7　Double-bus Wiring

The main advantages of double-bus wiring include reliable power supply, flexible operation, convenient maintenance and easy expansion, and it is widely used in large and medium-sized power plants and load-center substations.

2) Typical Operatoin

All the operations below are based on the example of Fig. 1-7.

(1) Section I bus operation to maintenance operation.

① Normal operation mode: Two sets of buses operate in parallel. L1, L3 and 5QF are connected to the Section I bus, and L2, L4 and 6QF are connected to the Section II bus.

Operation steps:

Verify that 0QF operates in the closing state, remove the operating power supply fuse 0QF, switch on the 52QS, disconnect the 51QS, switch on the 12QS, disconnect the 11QS, switch on the 32QS, disconnect the 31QS and put in use of the operating power supply fuse 0QF. Then disconnect the 0QF, check that 0QF has been disconnected, disconnect the 01QS and 02QS, shut down the voltage transformer of the Section I bus, take safety measures as required, and maintain the Section I bus. In the whole process, there is no circuit power outage.

In this process, remove the operating power supply fuse 0QF before operating the knife switch so that the bus-tie circuit breaker 0QF will not trip during the operation, and reliable equipotential on both sides of the operated knife switch can be ensured, because if the bus-tie circuit breaker trips during the operation, it may cause the disconnector to be disconnected (closed) with load and thus result in an accident.

② Normal operation mode: Section I bus is the working bus, and Section II bus is the standby bus.

Operation steps:

Switch on the bus-tie disconnectors 01QS and 02QS in turn, then switch on the bus-tie circuit breaker 0QF, and charge the standby bus with the bus-tie circuit breaker to check whether the standby bus is in good condition. If there is a short circuit fault in the standby bus, the bus-tie circuit breaker will trip immediately. If the standby bus is in good condition, it will not trip after switching on the bus-tie circuit breaker.

Then remove the operating power supply fuse 0QF, switch on the 52QS, disconnect the 51QS, switch on the 62QS, disconnect the 61QS, switch on the 12QS, disconnect the 11QS, switch on the 22QS, disconnect the 21QS, switch on the 32QS, disconnect the 31QS, switch on the 42QS, disconnect the 41QS, and enable the operating power supply fuse 0QF. Because the bus-tie circuit breaker is connected to two sets of buses, switching on and disconnecting the above disconnectors in turn only transfer the current, and will not generate the arc.

Disconnect the bus-tie circuit breaker 0QF, and disconnect bus-tie disconnectors 01QS and 02QS in turn. Then, the Section II bus is switched to a working bus, and the Section I bus is switched to a standby bus. During the above operation, any circuit is not affected.

(2) 51QS disconnector maintenance.

Normal operation mode: Two sets of buses operate in parallel. L1, L3 and 5QF are connected to the Section I bus, and L2, L4 and 6QF are connected to the Section II bus.

Operation steps: Just switch the lines L1 and L3 to the Section II bus for operation, then disconnect the circuit and the Section I bus connected with the disconnector, take safety measures, and the disconnector can be subject to interruption maintenance. For specific operation steps, please refer to Operation Section I bus operation to maintenance operation.

(3) The L1 circuit breaker 1QF refuses to operate.

The normal operation mode of cutting off the line L1 by the bus-tie circuit breaker is as follows: Two sets of buses operate in parallel; L1, L3 and 5QF are connected to the Section I bus, and L2, L4 and 6QF are connected to the Section II bus.

Operation steps:

Firstly, the circuit L3 and circuit 5QF are transferred from the Section I bus to the Section II bus for operation by bus transfer. At this time, the line L1, 1QF, Section I bus, bus-tie and Section II bus form a series-connected power supply circuit, and then the bus-tie circuit breaker 0QF is disconnected to cut off the circuit so as to ensure the reliable disconnection of line L1. For specific operation steps, readers can refer to the previous relevant operations to practice.

2. Sectionalized Double-bus Wiring

The sectionalized double-bus wiring is shown in Fig. 1-8. The Section I bus is divided into two sections by the section circuit breaker 00QF, and each section of bus is connected with the Section II bus through bus-tie circuit breakers 01QF and 02QF respectively. This wiring is more reliable and flexible than the double-bus wiring. When the I-bus is working and the II-bus is standby, it has the characteristics of single-bus sectional wiring. During maintenance of any section of Set I bus, the branch connected to this section of bus is transferred to the standby bus for operation, and the characteristics of single-bus sectionalized operation can still be maintained. When there are three or more power supplies, the power supplies can be respectively connected to the two sections of bus and the Set II bus, and the Set II bus is connected with one of sectionalized buses in Set I by the bus-tie circuit breaker to form a single bus running in three sections, which can further improve the reliability of power supply.

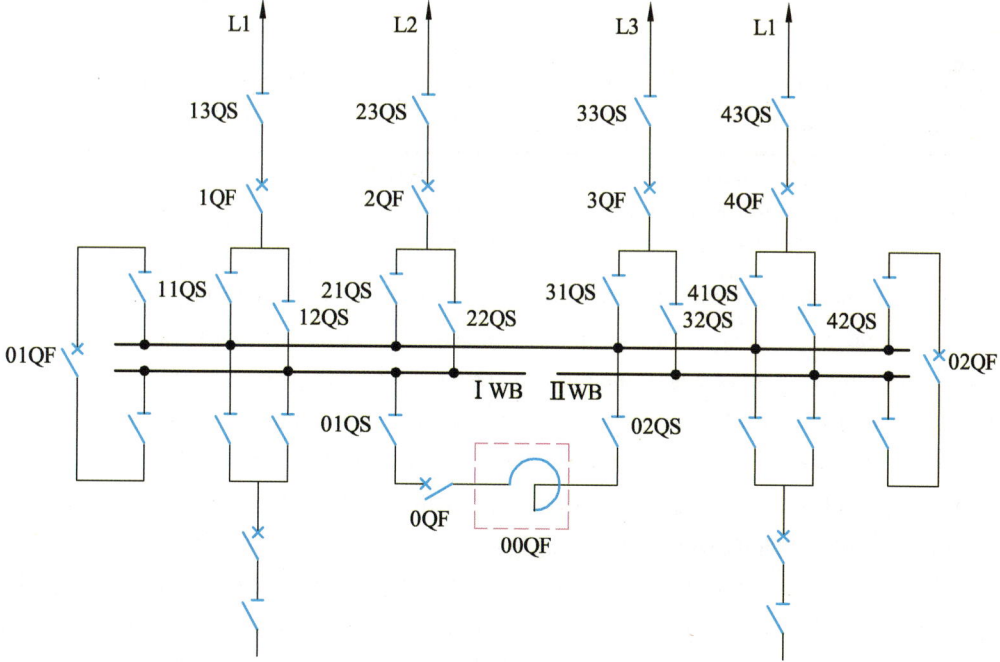

Fig. 1-8 Sectionalized Double-bus Wiring

Compared with the double-bus wiring, the sectionalized double-bus wiring increases two circuit breakers, which increases the investment. However, the double-bus section not only has various advantages of double-bus, but also has spare buses at any time, which has high reliability and flexibility.

3. Double-bus Wiring with Bypass Bus

1) Wiring Characteristics

The double-bus wiring with bypass and special bypass circuit breaker are shown in Fig. 1-9. The bypass circuit breaker can replace the outgoing circuit breaker, so that the power supply of the line is not affected during the maintenance of the outgoing circuit breaker. In normal operation of double-bus wiring with bypass, the fixed connection of two sets of buses is often adopted, that is, double buses run at the same time. At this time, the bus-tie circuit breaker is in the closed position, and some outgoing lines and power supplies are required to be fixed on the Section I bus, while the rest are connected to the Section II bus. To determine the fixed connection circuit of two sets of buses, both the reliability of power supply and the load balance should be considered to make the current passing through the bus-tie circuit breaker as small as possible.

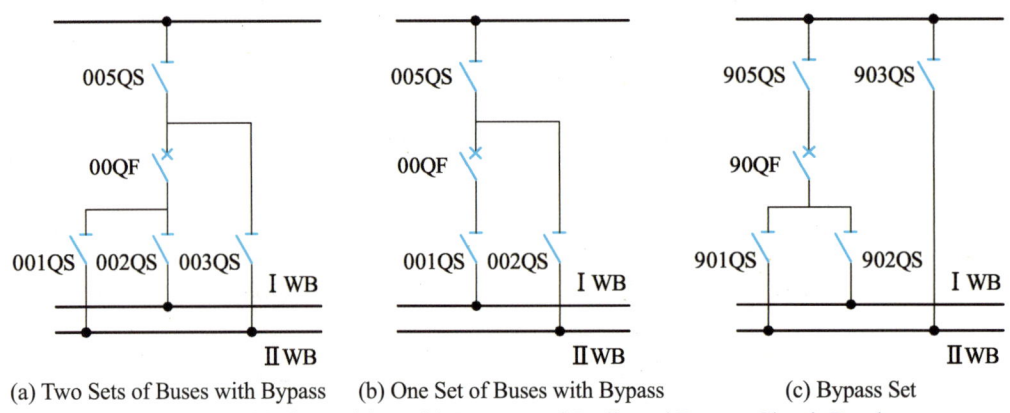

(a) Two Sets of Buses with Bypass　　(b) One Set of Buses with Bypass　　(c) Bypass Set

Fig. 1-9　Double-bus Wiring with Bypass and Dedicated Bypass Circuit Breaker

When the double-bus wiring with bypass is running in a fixed connection mode, a special bus differential protection device is usually provided. During operation, in the event of short-circuit fault in a set of buses, the bus protection device will trip the outgoing line, power supply and bus-tie circuit breaker connected with the bus to maintain the normal operation of the bus without fault. Then, according to the provisions of the operating procedures, the outgoing line and power supply circuit connected with the faulty bus can be switched to the bus without fault to resume power supply.

When a bypass circuit breaker is used to replace an outgoing circuit breaker for power supply, the bypass circuit breaker 90QF and the bus disconnector corresponding to the outgoing line should be switched on to maintain the original fixed connection mode.

When the number of outgoing lines is small and the utilization rate of the special bypass circuit breaker is not high, in order to save funds, the bus-tie circuit breaker can also be used as the wiring of the bypass circuit breaker. The specific wiring is shown in Fig. 1-10 (a), (b) and (c).

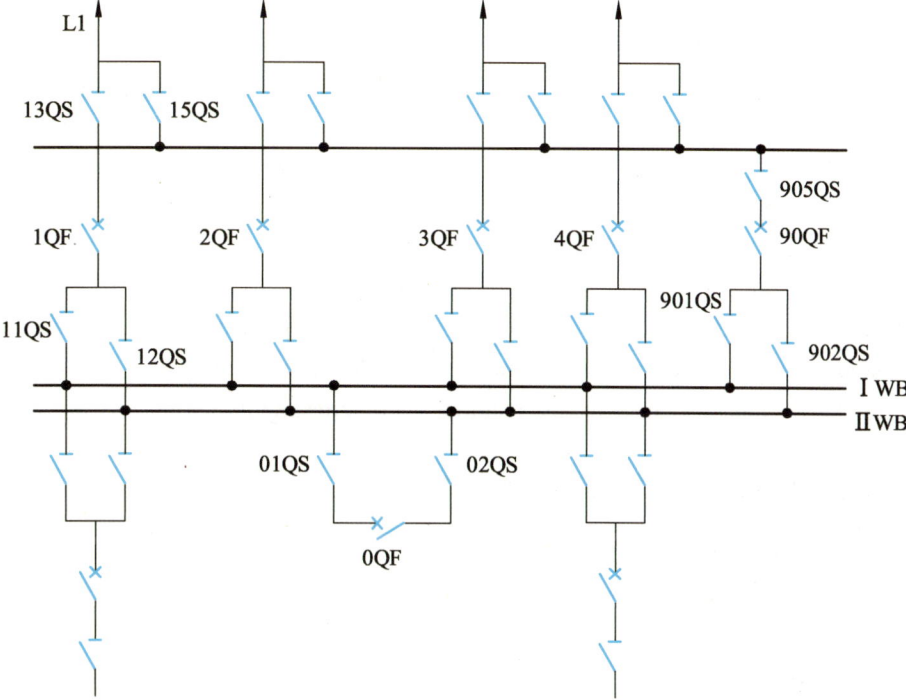

Fig. 1-10   Wiring of Bus-tie Circuit Breaker also Used as Bypass Circuit Breaker

For the wiring shown in Fig. 1-10 (a), during operation in the fixed connection mode, 002QS, 003QS and 00QF are closed, 001QS and 005QS are disconnected, the bypass bus is not live and the bypass circuit breaker 00QF operates as a bus-tie circuit breaker. If it is necessary to replace the outgoing circuit breaker with 00QF for power supply, the double-bus operation needs to be changed to single-bus operation, and then 00QF is used to replace the outgoing circuit breaker according to the operating procedures.

For the wiring shown in Fig. 1-10 (b), during operation in the fixed connection mode, 001QS, 00QF and 002QS are closed, 005QS is disconnected, the bypass bus is not live. When 00QF is used to replace the outgoing circuit breaker for power supply, it is necessary to switch the Section II bus to the standby bus, switch the Section I bus to the working bus and then use 00QF to replace the outgoing circuit breaker.

For the wiring shown in Fig. 1-10 (c), during operation in the fixed connection mode, 902QS, 90QF, 905QS and 903QS are closed, 901QS is disconnected, the bypass bus is live. When 90QF is used to replace the outgoing circuit breaker power supply, the double-bus operation needs to be changed to single-bus operation, and then 90QF is used to replace the outgoing circuit breaker according to the operating procedures.

2) Principle of Bypass Bus Setting

In the HV power distribution unit of 110 kV and above, due to the high voltage class, large transmission power and long transmission distance, the power outage has a great impact. Besides, it usually takes 5—7 days for maintenance of each high-voltage circuit breaker. Therefore, long-term power outage due to the maintenance of the circuit breaker is allowed, and it is necessary to set a bypass bus, so that when maintaining any circuit breaker connected to it, the circuit may not be power-off, thus improving the reliability of power supply. When the 110 kV outgoing line supplies for 6 circuits or more, and the 220 kV outgoing line supplies 4 circuits or more, the bypass bus with special bypass circuit breaker should be used.

With the widespread use of SF_6 circuit breakers in HV power distribution units, the quality improvement of domestic circuit breakers, the increase of system reserve capacity, the rationalization and close connection of power grid structure, the perfection of dual protection and the gradual transition from scheduled maintenance to condition-based equipment maintenance, the general trend is to gradually cancel bypass facilities in order to simplify wiring.

4. One-and-a-half Circuit Breaker Wiring

1) Wiring Characteristics

One-and-a-half circuit breaker wiring is shown in Fig. 1-11. There are two sets of buses. Each circuit is connected to a set of bus by a circuit breaker, and two circuits are connected

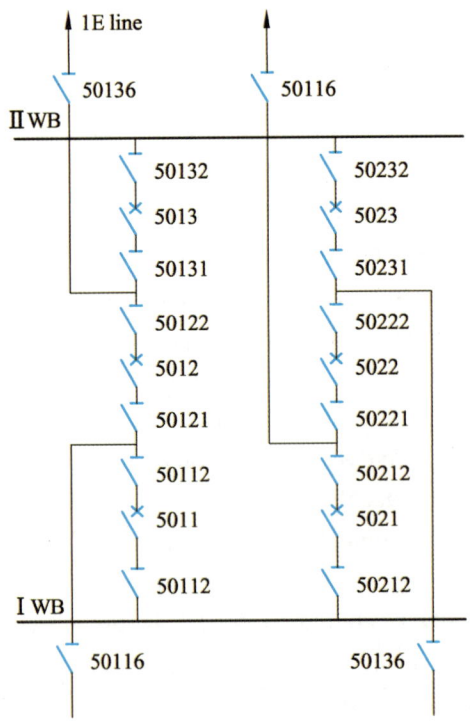

Fig. 1-11 One-and-a-half Circuit Breaker Wiring

by a circuit breaker, forming a string. All incoming and outgoing lines are connected to two circuit breakers, while three circuit breakers are used by two incoming and outgoing lines in the same string. Hence, one-and-a-half circuit breaker wiring is also called 3/2 wiring.

During normal operation, two sets of buses work at the same time, and all circuit breakers are closed, which is called complete series operation and forms multi-loop power supply, with very high reliability. Its main feature is that any bus fault or maintenance will not cause power outage; The maintenance of any circuit breaker will not cause power outage, either; Even in the extreme case where the two sets of buses fail at the same time, power can still be transmitted. When any circuit breaker in a string exits or is under maintenance, this method is called incomplete string operation, and it still does not affect the operation of any component. This kind of wiring is convenient to operate and simple to operate, and the disconnector is only used as isolated live equipment during maintenance.

2) Typical Operation

(1) Switch the Section I bus to maintenance.

① Disconnect the 5011 circuit breaker, and check that the 5011 circuit breaker is in the opening position;

② Disconnect the 5021 circuit breaker, and check that the 5021 circuit breaker is in the opening position;

③ Disconnect the 50111 disconnector, and check that the 50111 disconnector is in the opening position;

④ Disconnect the 50211 disconnector, and check that the 50211 disconnector is in the opening position;

⑤ After protection and safety measures are taken, the Section I bus can be maintained.

(2) Switch the 1E line from maintenance to operation.

① Withdraw safety and protection measures;

② Check that the 5012 circuit breaker is indeed disconnected;

③ Check that the 5013 circuit breaker is indeed disconnected;

④ Switch on the 50136 disconnector, and check that the 50136 disconnector is in the closed position;

⑤ Switch on the 5013 circuit breaker, and check that the 5013 circuit breaker is in the closed position;

⑥ Switch on the 5012 circuit breaker, and check that the 5012 circuit breaker is in the closed position;

(3) Switch the 5012 circuit breaker from operation to maintenance.

① Check that the 5012 circuit breaker is indeed disconnected;

② Disconnect the 50122 disconnector, and check that the 50122 disconnector is in the opening position;

③ Disconnect the 50121 disconnector, and check that the 50121 disconnector is in the opening position;

④ After protection and safety measures are taken, the 5012 circuit breaker can be maintained.

3) Scope of Application

The one-and-a-half circuit breaker wiring has been widely applied in 330~500 kV power distribution units of large power plants and transformer substations at home and abroad. When the number of incoming and outgoing circuits is 6 or more, and they play an important role in the system, it is advisable to use one-and-a-half circuit breaker wiring.

5. 4/3 Circuit Breaker Wiring

Due to the high cost of high-voltage circuit breakers, in order to further reduce the equipment investment, the incoming and outgoing lines of 3 circuits are connected to two sets of buses through 4 circuit breakers to form a 4/3 circuit breaker wiring mode, as shown in Fig. 1-12. This wiring mode is usually used in large hydropower plants where the number of generators (incoming lines) is greater than the number of lines (outgoing lines), so as to realize the matching of power supply and load capacity in the 3 circuits of a string.

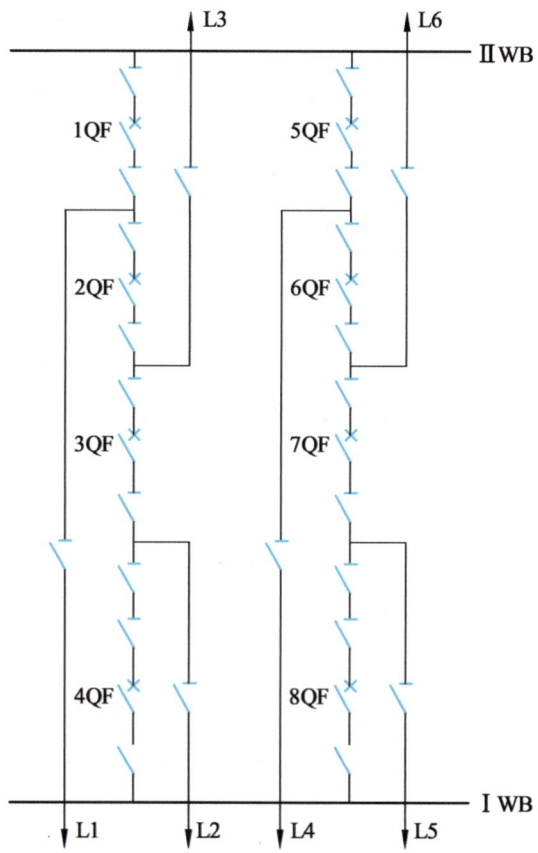

Fig. 1-12 4/3 Circuit Breaker Wiring

In practical application, according to the quantity of power supply and load and expansion requirements, the combined wiring of 4/3, one-and-a-half and two circuit breakers can be adopted, which will help to improve the reliability and flexibility of power distribution units.

6. Wiring of Transformer Bus Set

In addition to the above common kinds of wiring, the wiring of transformer bus set as shown in Fig. 1-13 can also be used. This wiring transformer is directly connected to the bus, and each outgoing circuit is connected by double circuit breakers, as shown in Fig. 3-13 (a) or by one-and-a half circuit breakers, as shown in Fig. 1-13 (b). This mode has the advantages of flexible scheduling, free allocation of power supply and load, safety and reliability, and expansion convenience.

Because of the high operation reliability of the transformer, the direct connection to the bus has no obvious influence on the operation of the bus. Once the transformer fails, the circuit breaker connected to the bus will trip, but it will not affect the power supply of other circuits. After the failed transformer is withdrawn with the disconnector, the bus can resume operation through switching operation. This kind of wiring is suitable for transformer substations with long-distance large-capacity transmission lines, outstanding system stability problems, high line reliability of lines requirements, and requirements of reliable quality of main transformers and low failure rate.

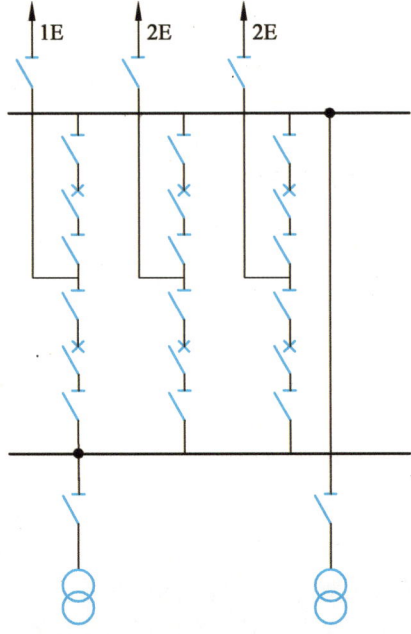

(a) Double Circuit Breaker Wiring of Outgoing Line

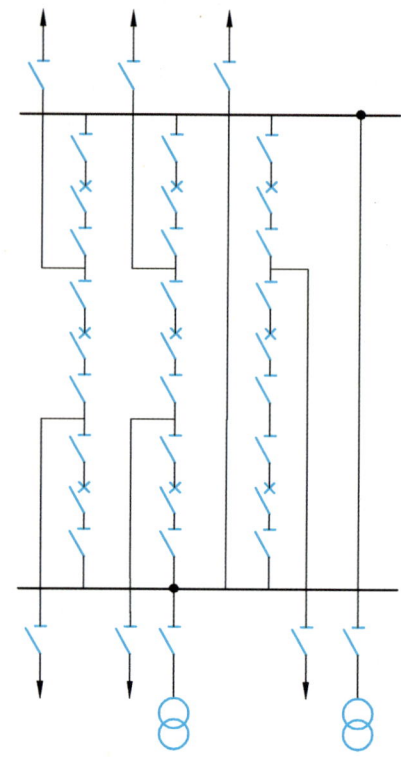

(b) One-and-a-half Circuit Breaker Wiring of Outgoing Line

Fig. 1-13　Wiring of Transformer Bus Set

模块四　无母线接线

一、单元接线及扩大单元接线

单元接线是将不同的电气设备（发电机、变压器、线路）串联成一个整体，称为一个单元，然后再与其他单元并列。

1. 单元接线

单元接线如图1-14所示。图1-14（a）为发电机—双绕组变压器组成的单元，断路器装于主变高压侧作为该单元共同的操作和保护电器，在发电机和变压器之间不设断路器，可装一组隔离开关供试验和检修时作为隔离元件。

当高压侧需要联按两个电压等级时，主变压器采用三绕组变压器或自耦变压器，就组成发电机—三绕组变压器（自耦变压器）单元接线，如图1-14（b）、（c）所示。为了能保证发电机故障或检修时高压侧与中压侧之间的联系，应在发电机与变压器之间装设断路器。若高压侧与中压侧对侧无电源时，发电机和变压器之间可不设断路器。

图1-14（d）为发电机－变压器－线路组单元接线。它是将发电机、变压器和线路直接串联，中间除了自用电外没有其他分支引出。这种接线实际上是发电机—变压器单

元和变压器－线路单元的组合，常用于 1~2 台发电机、一回输电线路，且不带近区负荷的梯级开发的水电站，把电能送到梯级开发的联合开关站。

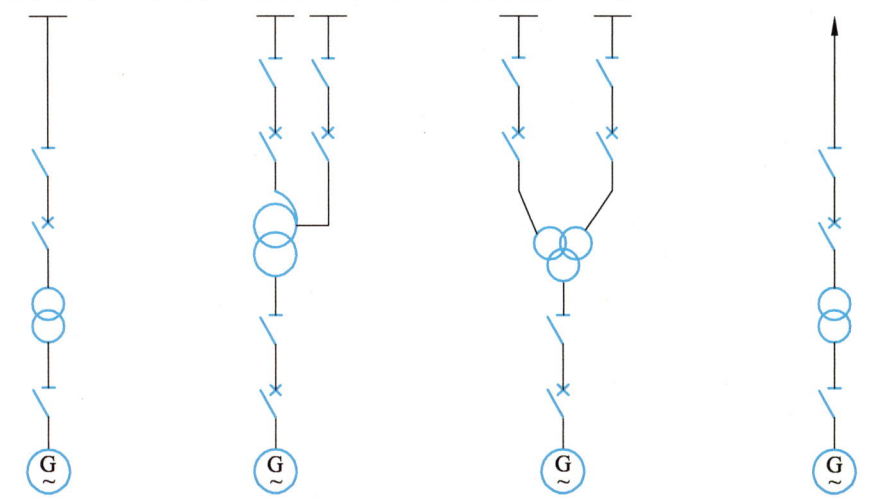

(a) 发电机双绕组变压器 (b) 发电机自耦变压器单 (c) 发电机三绕组变压器单元 (d) 发电机—变压器—线
单元接线　　　　　　　元接线　　　　　　　　接线　　　　　　　　　路组单元接线

图 1-14　单元接线

2. 扩大单元接线

采用两台发电机与一台变压器组成单元的接线称为扩大单元接线，如图 1-15 所示。在这种接线中，为了适应机组开停的需要，每一台发电机回路都装设断路器，并在每台发电机与变压器之间装设隔离开关，以保证停机检修的安全。装设发电机出口断路器的目的是使两台发电机可以分别投入运行或当任一台发电机需要停止运行或发生故障时，可以操作该断路器，而不影响另一台发电机与变压器的正常运行。

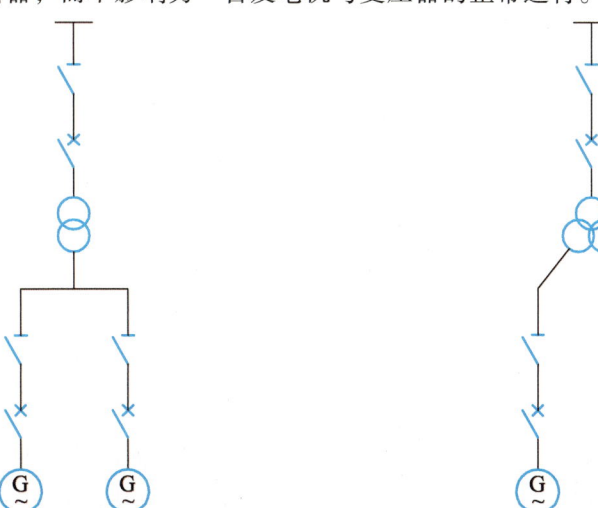

(a) 发电机双绕组变压器扩大单元接线　　　　(b) 发电机分裂绕组变压器扩大单元接线

图 1-15　扩大单元接线

扩大单元接线与单元接线相比有如下特点：

（1）减小了主变压器和主变高压侧断路器的数量，减少了高压侧接线的回路数，从而简化了高压侧接线，节省了投资和场地。

（2）任一台机组停机都不影响厂用电的供给。

（3）当变压器发生故障或检修时，该单元的所有发电机都将无法运行。

扩大单元接线用于在系统有备用容量时的大中型发电厂中。

二、桥形接线

桥形接线适用于仅有两台变压器和两回出线的装置中，接线如图 1-16 所示。桥形接线仅用三台断路器，根据桥回路（3QF）的安装位置不同，可分为内桥和外桥两种接线。桥形接线正常运行时，三台断路器均闭合工作。

（一）内桥接线

内桥接线如图 1-16（a）所示，桥回路置于线路断路器内侧（靠变压器侧），此时线路经断路器和隔离开关接至桥接点，构成独立单元；而变压器支路只经隔离开关与桥接点相连，是非独立单元。

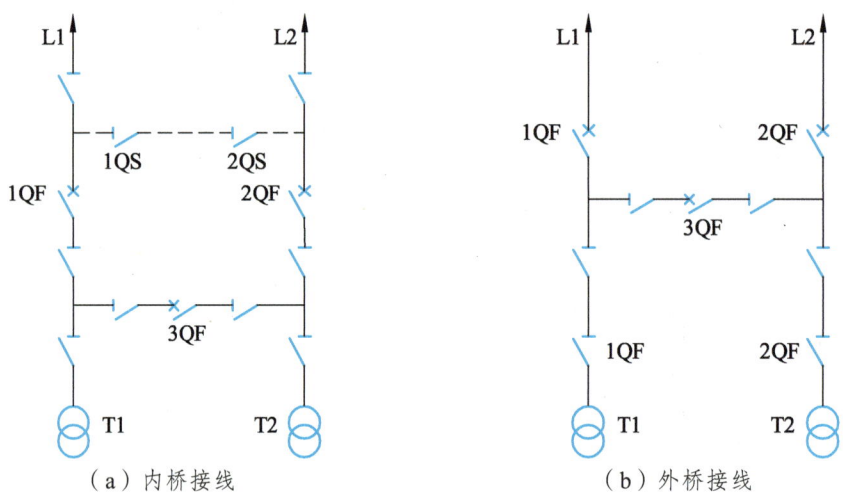

（a）内桥接线　　　　　　　　（b）外桥接线

图 1-16　桥形接线

内桥接线的特点为：

（1）线路操作方便。如线路发生故障，仅故障线路的断路器跳闸，其余三回路可继续工作，并保持相互的联系。

（2）正常运行时变压器操作复杂。如变压器 T1 检修或发生故障时，需断开断路器 1QF、3QF，使无故障线路 L1 供电受到影响；然后需经倒闸操作，拉开隔离开关 1QS 后，再合上 1QF、3QF 才能恢复线路 L1 工作。因此，将造成该侧线路的短时停电。

（3）桥回路故障或检修时两个单元之间失去联系；同时，出线断路器故障或检修时，造成该回路停电。为此，在实际接线中可采用设外跨条来提高运行灵活性。

（二）外桥接线

外桥接线如图 1-16（b）所示。桥回路置于线路断路器外侧，变压器经断路器和隔离开关接至桥接点，而线路支路只经隔离开关与桥接点相连。

外桥接线的特点为：

（1）变压器操作方便。如变压器发生故障时，仅故障变压器回路的断路器自动跳闸，其余三回路可继续工作，并保持相互的联系。

（2）线路投入与切除时操作复杂。如线路检修或故障时，需断开两台断路器，并使该侧变压器停止运行，需经倒闸操作恢复变压器工作，造成变压器短时停电。

（3）桥回路故障或检修时，两个单元之间失去联系，出现侧断路器故障或检修时，造成该侧变压器停电，在实际接线中可采用设内跨条来解决这个问题。

三、多角形接线

多角形接线也称为多边形接线，如图 1-17 所示。它相当于将单母线按电源和出线数目分段，然后连接成一个环形的接线。比较常用的有三角形、四角形和五角形接线。

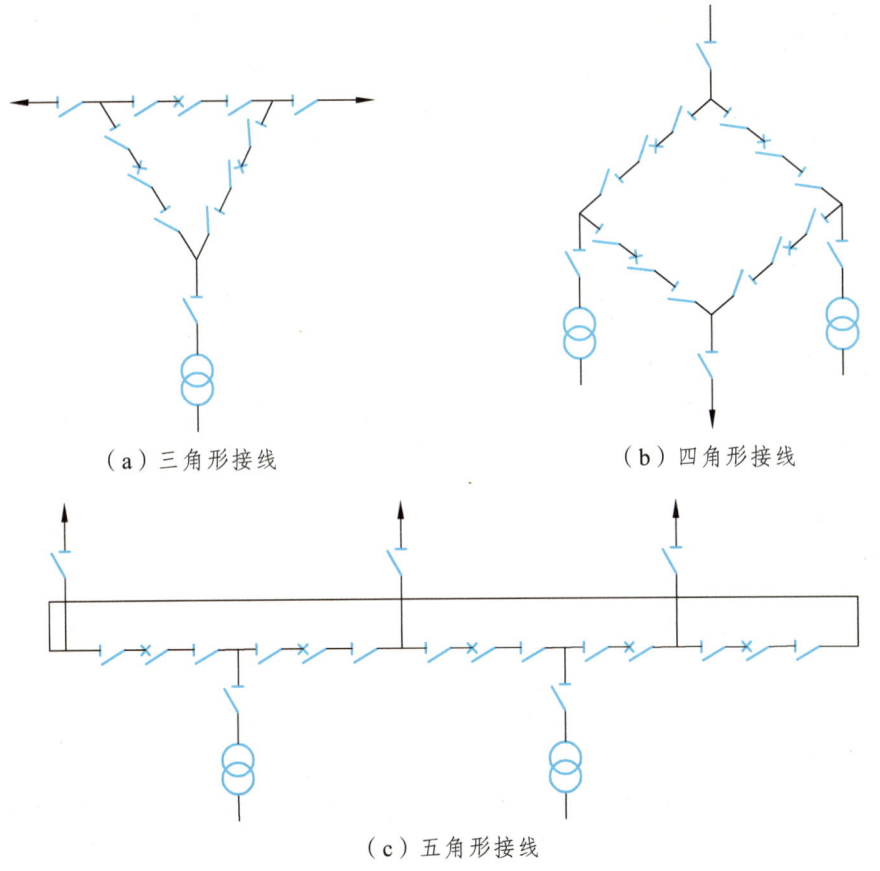

（a）三角形接线　　　　（b）四角形接线

（c）五角形接线

图 1-17　多角形接线

多角形接线具有如下特点：

（1）每个回路位于两个断路器之间，具有双断路器接线的优点，检修任一断路器都不中断供电。

（2）所有隔离开关只作隔离电器用，不作操作电器用，容易实现自动化和遥控。

（3）正常运行时，多角形是闭合的，任一进出线回路发生故障，仅该回路断开，其余回路不受影响，因此运行可靠性高。

（4）任一断路器故障或检修时，则开环运行，此时若环上某一元件再发生故障就有可能出现非故障回路被迫切除并将系统解列。这种缺点随角数的增加更为突出，所以这种接线最多不超过 6 角。

（5）开环和闭环运行时，流过断路器的工作电流不同，这将给设备选择和继电保护整定带来一定的困难。

（6）此接线的配电装置不便于扩建和发展。

Module 4　Bus-free Wiring

1. Unit Wiring

Unit wiring is to connect different electrical equipments (generators, transformers and lines) into a whole, which is called a unit, and then connect in parallel with other units.

1) Unit Wiring

The unit wiring is shown in Fig. 1-14. Fig. 1-14 (a) shows a unit consisting of a generator and a double-winding transformer. The circuit breaker is installed on the HV side of the main transformer as a common operation and protection appliance of the unit. There is no circuit breaker between the generator and the transformer, and a set of disconnectors can be installed as isolation elements for testing and maintenance.

When the HV side needs to connect two voltage classes, the main transformer adopts a three-winding transformer or an autotransformer to form the unit wiring of generator—three-winding transformer (autotransformer), as shown in Fig. 1-14 (b) and (c). In order to ensure the connection between the HV side and the MV side when the generator fails or is under maintenance, a circuit breaker should be installed between the generator and the transformer. If there is no power supply between the HV side and the MV side, there is no need to install a circuit breaker between the generator and the transformer.

Fig. 1-14 (d) shows the unit wiring of generator—transformer—line set. It connects the generator, transformer and line in series, and there is no other branch except for non-utility power supply. This kind of connection is actually a combination of generator-transformer unit and transformer-line unit, which is often used in hydropower stations of cascade development with 1—2 generators and primary transmission lines and without near-area load, to transmit electric energy to the joint switch station of cascade development.

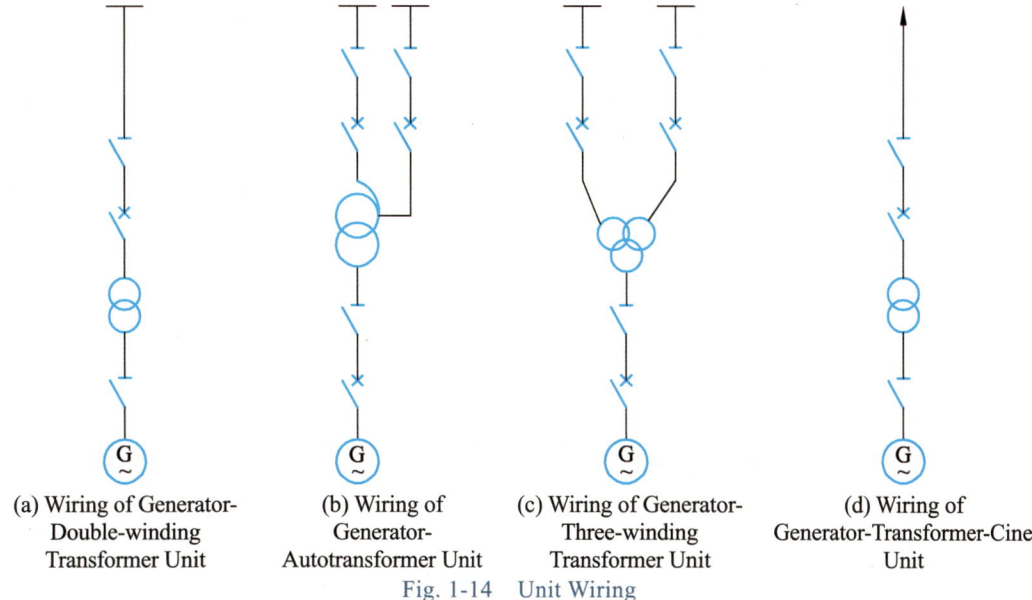

Fig. 1-14 Unit Wiring

(a) Wiring of Generator-Double-winding Transformer Unit
(b) Wiring of Generator-Autotransformer Unit
(c) Wiring of Generator-Three-winding Transformer Unit
(d) Wiring of Generator-Transformer-Cine Unit

2) Expanded Unit Wiring

The unit wiring consisting of two generators and a transformer is called expanded unit wiring, as shown in Fig. 1-15. In this wiring, in order to meet the requirements of unit startup and shutdown, each generator circuit is equipped with a circuit breaker, and a disconnector is installed between each generator and transformer to ensure the safety of shutdown and maintenance. The purpose of installing the generator outlet circuit breaker is to enable two generators separately or to operate the circuit breaker when either generator needs to stop running or fails, without affecting the normal operation of the other generator and transformer.

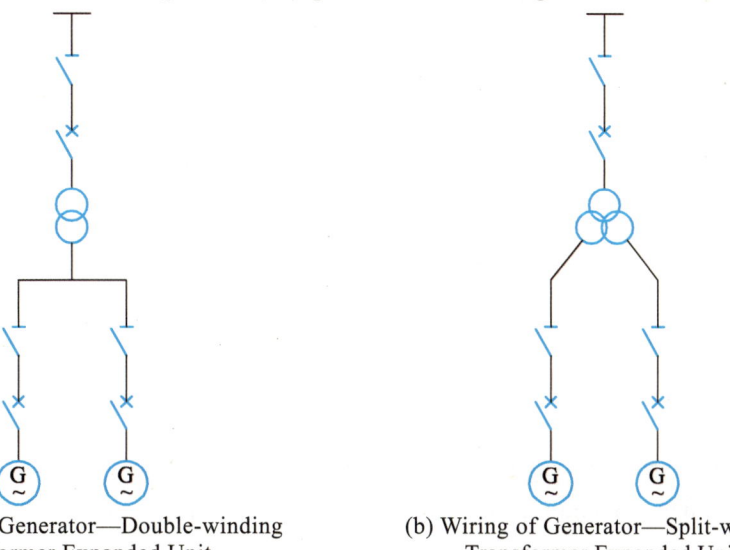

(a) Wiring of Generator—Double-winding Transformer Expanded Unit
(b) Wiring of Generator—Split-winding Transformer Expanded Unit

Fig. 1-15 Expanded Unit Wiring

Compared with the unit wiring, the expanded unit wiring has the following characteristics:

(1) It has reduced the number of main transformers and the number of circuit breakers at the HV side of main transformer, and decreased the number of circuits for the wiring at the HV side, thereby simplifying the wiring at the HV side and saving the investment and site.

(2) The supply of service power will not be influenced by the shutdown of any unit.

(3) All generators of this unit cannot run in case of any fault or maintenance of the transformer.

Expanded unit wiring is used in large and medium-sized power plants with backup capacity of the system.

2. Bridge Wiring

Bridge wiring is applicable to the devices with only two transformers and two outgoing lines (shown in Fig. 1-16). Only three circuit breakers are used in the bridge wiring. According to the different installation positions of the bridge circuit (3QF), bridge wiring can be divided into two types of wiring: inner bridge and outer bridge. When the bridge wiring is in normal operation, all the three circuit breakers are closed.

1) Internal Bridge Wiring

The internal bridge wiring is shown in Fig. 1-16 (a), and the bridge circuit is placed inside the circuit breaker (near the transformer side). At this time, the line is connected to the bridge point through the circuit breaker and disconnector, forming an independent unit. The transformer branch is only connected to the bridge point through the disconnector, which is a non-independent unit.

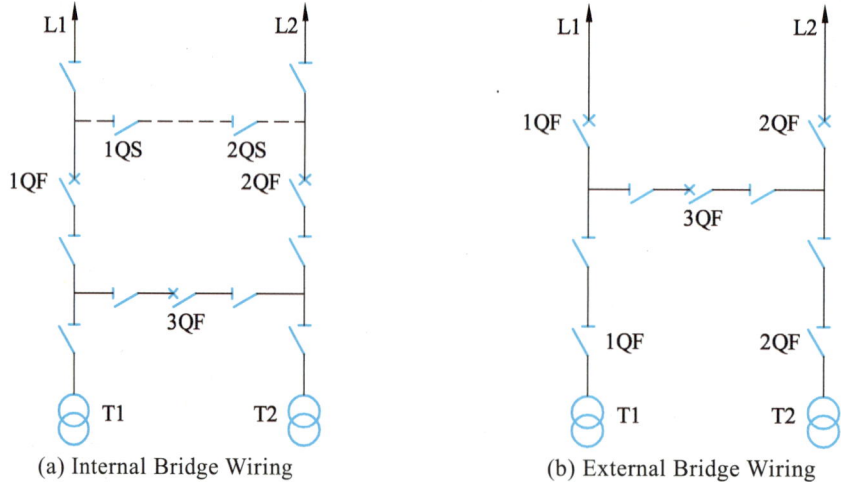

(a) Internal Bridge Wiring (b) External Bridge Wiring

Fig. 1-16 Bridge Wiring

The characteristics of internal bridge wiring are as follows:

(1) Easy line operation. In the event of line fault, only the circuit breaker of the faulty line trips, and the other three circuits can continue to work and keep in touch with each other.

(2) Complicated transformer operation during normal operation. If the transformer T1 is

under maintenance or fails, it is necessary to disconnect the circuit breakers 1QF and 3QF, which affects the power supply for the non-faulty line L1. Then, switching operation is required. After the disconnector 1QS is disconnected, then the circuit breakers 1QF and 3QF are switched on to restore the line L1. Therefore, it will cause a short-term power outage of the line on this side.

(3) When the bridge circuit fails or is under maintenance, the two units lose contact. In addition, the power outage is caused to the circuit when the outgoing circuit breaker fails or is under maintenance. Hence, external cross-connectors can be used in the actual wiring to improve operational flexibility.

2) External Bridge Wiring

The external bridge wiring is shown in Fig. 1-16 (b), and the bridge circuit is placed outside the circuit breaker. The transformer is connected to the bridge point through the circuit breaker and disconnector, while the line branch is connected to the bridge point only through the disconnector.

The characteristics of external bridge wiring are as follows:

(1) Easy transformer operation. In the event of transformer fault, only the circuit breaker of the faulty transformer circuit trips, and the other three circuits can continue to work and keep in touch with each other.

(2) Complicated operation when the line is put into operation and cut off. In case of line maintenance or fault, two circuit breakers needs to be disconnected and the transformer on this side should stop running. It is necessary to restore the operation of the transformer through switching operation, thus resulting in short-term power outage of the transformer.

(3) In case of any bridge circuit fault or maintenance, the two units lose contact. When the circuit breaker on the side of the outgoing line fails or is under maintenance, the transformer on this side is power-off. In actual wiring, this problem can be solved by setting internal cross-connectors.

3. Polygonal Wiring

Polygonal wiring is shown in Fig. 1-17. It is equivalent to segmentation of a single bus according to the number of power sources and outgoing lines, and then connecting them into a ring-shaped wiring. Common polygonal wiring includes triangular, tetragonal and pentagonal wiring.

Polygonal wiring has the following characteristics:

(1) Each circuit is located between two circuit breakers, with the advantage of double circuit breaker wiring, and the power supply will not be interrupted during maintenance of any circuit breaker.

(2) All disconnectors are only used as isolating electrical appliances rather than operating electrical appliances, which is easy to realize automation and remote control.

(3) During normal operation, the polygon is closed. When any incoming and outgoing circuit fails, only this circuit is disconnected, and the rest circuits are not affected. Therefore, the operation reliability is high.

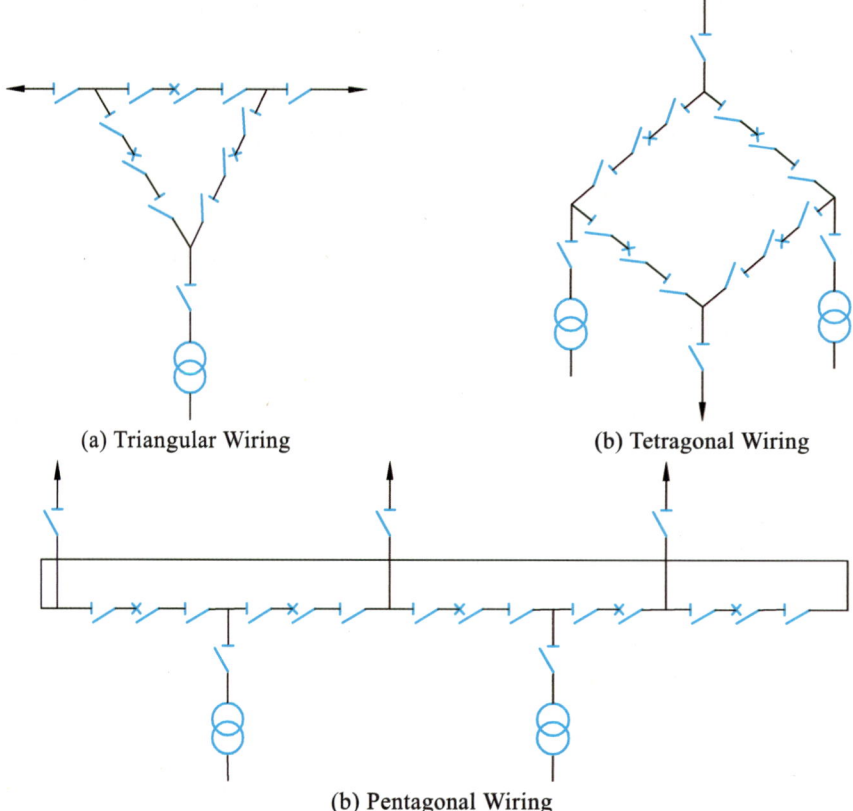

(a) Triangular Wiring   (b) Tetragonal Wiring

(b) Pentagonal Wiring

Fig. 1-17　Polygonal Wiring

(4) When any circuit breaker fails or is under maintenance, it will operate in the open loop. At this time, if a component on the loop fails again, the non-fault circuit may be forced to cut off and the system will be disconnected. This disadvantage is more prominent with the increase in the number of angles, so the number of angles should not exceed 6 in this kind of wiring.

(5) In open-loop and closed-loop operation, the working current flowing through the circuit breaker differs, which will bring some difficulties to equipment selection and relay protection setting.

(6) The power distribution unit of this wiring is not convenient for expansion and development.

任务一　变电站现场主接线认知

一、工作任务

110 kV 智能变电站主接线认知，掌握变电站电气一次主接线的主要形式，进一步理解母线设备在电力系统中的应用。

二、工作要求

（1）作业人员精神状态良好，熟悉工作中安全措施、技术措施以及现场工作危险点。
（2）实训现场要求按生产现场规范布置安全措施，并严格执行标准化作业。
（3）作业人员应规范穿戴工作服和安全帽，做好安全防护。

三、工作准备

危险点及预控措施见表 1-1。

表 1-1　危险点及预控措施表

序号	防范类型	危险点	预防控制措施
1	误碰、误动、误登电气设备	意外伤人	观察设备时应与带电设备保持足够的安全距离。110 kV 不小于 1.5 m，35 kV 不小于 1.0 m，10 kV 不小于 0.70 m
2	擅自打开设备网门，擅自移动临时安全围栏，擅自跨越设备固定围栏	观察设备低压触电	观察时，不得进行其它操作（严禁进行电气操作），不得移开或越过遮栏
3	擅自改变现场设备状态，变更工地点安全措施	观察人员误碰、误登带电设备	观查设备禁止变更检修现场安全措施，禁止改变检修设备状态
4	特殊天气未按安全规定佩戴安全防护用具	意外伤人	应按规程佩戴安全防护用具
5	雷雨天气，靠近避雷器和避雷针，造成人员伤亡	意外伤人	雷雨天气，需要巡视高压设备时，应穿绝缘靴，并不得靠近避雷器和避雷针
6	进出高压室，未随手关门，造成小动物进入	小动物进入短路	进出高压室，必须随手将门锁好
7	不戴安全帽、不按规定着装，在突发事件时失去保护	高空落物伤人	进入设备区，必须戴安全帽
8	现场安全措施不规范，如警告标识不齐全、孔洞封闭不良、带电设备隔离不符合要求原因，易造成人身伤害	工作所需各类生产、安全工器具、资料等不能满足现场工作需求，导致人员触电	使用前进行全面检查，不合格工器具禁止带入工作现场

四、作业流程

作业流程见表 1-2。

表 1-2　作业流程表

序号	作业内容	作业步骤及标准
1	工作前准备工作	检查着装和安全帽是否穿戴合格；交代危险点、安全措施；总结每小组将观查到的站内设备连接方式写成报告
2	熟悉 110 kV 智能变电站主接线图	交代本次工作任务
3	认知 110 kV 主接线	分组在变电站里进行主接线上的设备对应，观察设备间的连接方式

Task 1　Cognition of Main Wiring in Transformer Substation

1. Work Tasks

Learn about the main wiring of 110 kV intelligent substation, master the main forms of electrical primary main wiring of the substation, and further understand the application of bus equipment in the power system.

2. Work Requirements

(1) The operators shall be in a good mental state, and familiar with safety measures, technical measures and on-site hazards.

(2) Arrange safety measures on practical training site according to the production site specification, and strictly implement standard operation.

(3) The operators shall wear working clothes and safety helmet and take safety protection measures according to the specification.

3. Preparation for Work

Hazards and preventive and control measures are shown in Table 1-1.

Table 1-1　Hazards and Preventive and Control Measures

S/N	Type of precautions	Hazard	Preventive and control measures
1	Don't touch, operate or climb on the operating equipment.	Accidental personal injury.	Keep enough safe distance from live equipment when observing the equipment. The distance shall be no less than 1.5 m for 110 kV equipment, no less than 1.00 m for 35 kV equipment, and no less than 0.70 m for 10 kV equipment.
2	Open the equipment door, move temporary safety fence, or stride over the fixed equipment fence without authorization.	Low-voltage electric shock when the operator is observing the equipment.	During observation, don't engage in other work (electrical work is strictly prohibited), move or cross the fence.
3	Change the site equipment state and the safety measures at the workplace without authorization.	The observer touches or climbs on live equipment accidentally.	Do not change safety measures on the maintenance site or the equipment maintenance state when the observer is observing the equipment.
4	Fail to wear safety protection equipment according to safety regulations in special weather.	Accidental personal injury.	Wear safety protection equipment according to safety regulations.
5	Be close to lightning arrester and lightning rod in the thunderstorm weather, causing casualties.	Accidental personal injury.	In the thunderstorm weather, when it is necessary to inspect outdoor high-voltage equipment, the insulating boots shall be worn, and the lightning arresters and lightning rods shall not be approached.
6	Fail to close the door after entering and going out from the high-voltage room, causing small animals to enter the room.	Enterance of small animals into the high-voltage room, causing short circuit.	Close the door after entering and going out from the high-voltage room.
7	Fail to wear safety helmet and working clothes as specified, so that protective measures cannot be provided in case of emergency.	Personal injury caused by falling objects.	The operator shall wear safety helmet when entering the equipment area.
8	The safety measures on site do not meet the requirements. For example, warning signs are not complete, holes are poorly closed, and the isolation of live equipment does not meet the requirements, which may easily cause personal injury.	The production and safety tools and instruments and data required for work cannot meet the job requirements on site, causing electric shock.	Conduct a comprehensive inspection before use and do not bring non-conforming tools and instruments to the work site.

项目一　电气主接线

4. Operation Procedures

The operation procedures are shown in Table 1-2.

Table 1-2　Operation Procedures

S/N	Operation content	Operational steps and standards
1	Preparations before work	1. Check whether the dress and safety clothing are qualified; 2. Explain this task; 3. Inform the hazards and safety measures.
2	Learn about 110 kV main wiring	1. Know well the main electrical wiring diagram of 110 kV intelligent substation; 2. Group to correspond the devices on the main wiring in the substation, and observe the connection mode between the devices.
3	Summary	Each group submits the report on the observed equipment connection mode in the substation.

项目二　电气设备检修标准化作业

在电力设备的检修现场生产中，标准化作业能提供的最佳结果就是作业现场有条不紊的秩序。标准化在生产中的标志就是流程化、指令性、可替代、闭环可追溯、质量控制，尽量排除个体能力导致的结果差异。

在电力系统生产现场管控体系中，任何专业的生产标准化作业，都体现为现场作业进行秩序管控，它的依据和管控对象主要来自安全生产和技术活动。将反复进行的相同工作，按照"完美工人"原则进行，将全过程中的各项步骤分解为动作细节和对应的指标要素，依据各要素在操作执行中必须依循的客观规律要求一一罗列了。并按照一定的行为和思维逻辑顺序排列这些要素，将其组合成为更合理的作业程序和步骤，并形成文档形式的标准化模板资料发放给各实施单位。在工作实施前，施工人员需要将获得的标准化模板结合现场实际情况进行针对化编制形成最终的标准化作业文档，并上报企业审批。在检修工作实施中，施工人员的任务就是严格按照批准的标准化作业文档，不打折扣地进行施工。

应该认识到的是，标准化作业在尽量排除工作人员个人能力差异导致的作业结果的同时，客观上防范了安全事故的发生。这方面的作业和近些年来各行业提倡的"清单革命"非常类似，本章内容主要以变电设备检修为主。

模块一　现场危险点辨识

电气设备检修涉及的设备种类多、型号多、试验项目多、一二次设备混合、工作场地情况多变、外来人员共同施工等情况在工作中都会遇到，且变电检修绝多大数的工作都是由人来完成，由于个体技能、经验和认知水平的差异，导致哪怕是同一项工作换不同的员工来进行，对工作现场的危险点辨识也可能出现差异，这就可能导致工作中不安全情况的发生。本节内容旨在通过对危险点辨识的基本原则、要点的阐述，结合案例让读者对变电检修的一些典型工作危险点能有充分的认知。

一、危险点分析的要点

在变电检修工作开展中，由于具体的工作环境及相应的检修条件存在一定程度的差异，所以危险点存在的情况及具体的分析要点就会不尽相同。如果检修人员是在高空作业状态下进行检修，那么危险点的位置就相对较高，检修人员很容易出现高空坠落等。如果是在带电状态下进行检修，检修人员也会面临明显的带电危险。此外，在很多情况下，变电检修的环境中存在有害气体等，这时危险点就是有害气体等可能给检修人员造成的危害，也可能给检修设备等带来危害。对于变电检修工作而言，检修人员所使用的设备等也是危险点之一。

比如一些大型的机械设备，如起重机等，若是检修人员在操作过程中出现失误等，也可能造成检修人员受伤等，这也是危险点之一。所以检修人员应当认识到，一定要按照相应的流程及规范要求进行操作，从而杜绝安全事故的发生。在变电检修危险点分析的要点当中，检修人员还应当认识到自身心理安全意识的重要意义，并且针对存在的消极心理等进行及时的疏导等，不要存在明显的畏难心理等。检修人员需要保证在实际工作当中有充沛的精力及积极的心理，如果自身心理状况不佳，很容易在实际工作当中出现抗压性不强等问题，甚至因走神等出现安全事故。可参考附表1。

二、危险点分析的主要步骤

1. 总结经验，借鉴资料

在对变电检修中存在的危险点进行分析的时候，相关人员首先需要对危险点自身的性质、特点及主要产生的影响等进行充分考虑，并且做好可能出现的危险事故方面的预判处理。相关人员应当多对过去变电检修工作中出现过的事故问题等进行分析，并且总结具体的经验教训等，吸收优秀的经验，为变电检修工作提供一手资料。检修人员还可以针对此前出现的危险点资料等与现在的检修环境等相关因素进行分析，以此进行借鉴。

2. 落实到人，保持沟通

提前召开会议分工布置。对于检修当中可能出现的危险点，检修人员需要与其他工作人员共同分析，开展集中讨论，共同提出预防对策。做到每一个位置都有相应的负起责任的检测人员。最后汇总分析各个危险点的数据，进行统一指导工作。针对确定的危险点制定合理的安全保护措施，并将情况无隐瞒地告诉操作者。不论评估多准确都需要实际操作人员在具体的作业中进行实地检测。这项工作繁重而且复杂，且随时又会有新的状况发生，这就要求工作者与保护者进行密切沟通，发现问题上报，妥善合理地及时解决。

3. 分组管理，制定预案

检修人员需要制定好应急处理预案，根据危险点存在的类型等制定预案，并且明确每一名检修人员及其他工作人员的职责，将不同的工作人员分为不同小组，以便进行紧急处理等。

三、作业现场的安全控制

1. 加强现场安全工作的过程管理

从目前的安全工作流程来看，标准化的工作流程必须匹配标准化的现场安全措施。在每个作业过程中，着重明确所有参加工作的领导及相关工作人员的职责和分工，对每一个存在威胁设备、人身安全的环节都用标准化的流程来管理和执行。确保防患于未然，把危害降低到最小值。在萌芽中消除隐患。做到一切不良现象的及时发现和适当制止。

2. 检修现场作业实施

检修现场作业安全措施标准基本流程应包括现场勘查、标准编制、上级审核、标准实施、总结、评估及考核等关键环节。

（1）现场勘查：为提前检测作业可能潜在的危险，方便制定保护措施，要求作业单位事先抵达变电站进行现场实际勘探。安全措施要在所检测设备完全断电的前提之下，对可能来电的检修设备电侧装设接地线或合上接地刀闸，避免安全措施重复带来的经费浪费和人力、物力、财力的浪费。同时要兼顾考虑设备检修工作实际，保证检修的同时使作业现场有相应的保护措施。

（2）标准编制：依照实地勘查结果，制定图卡。通常需要把作业场地所需安全措施、潜在危险位置、相关安全措施详细填写在同一措施内。还需在变电站的平面图上详细注明现场设备的实际运行状态和安全设施布控。

（3）上级审核：上级严格、定期统一审批编制卡片，才能确保检修现场作业安全措施的具体实施落实到实处，具有一定的约束作用和强制作用。

（4）标准实施：必须严格按照批准实施的文件落实现场安全及危险点控制措施、相关人员的分内责任，确保突发状况发生之际分工明确、各司其职。

（5）总结、评估、考核：检修工作完成后立刻进行工作总结，对编制应用检修现场作业安全措施标准的效果进行评估，从工作中总结经验，完善安全措施流程。

3. 做好监测工作

电力单位及变电检修人员需要针对正在运行期的相关设备进行严格监测，确切了解设备目前的运行状态、安全性能等，帮助变电检修人员对设备的相关性能等进行调整，以此保证设备中不存在危险点。此外，及时淘汰一些过于老旧的设备，以免其在运行或检修期间出现安全事故等，特别是一旦失灵就可能造成人身伤害的非电量部件，如储能弹簧、密度继电器、密封装置、气体或油阀门、起吊用钢丝绳等，这些设备和物品状态几乎无法在电量保护装置上显示，因此需要加强运维监测，以提供检修工作的危险点判断依据。

Program 2 Standardized Operation of Electrical Equipment Maintenance

On the maintenance site of electrical equipment, the optimum result of standardized operation is that everything is put in good order and is well arranged on the work site. Standardized operation in production is represented by process management, command control, replaceability, closed-loop traceability, and quality control, in order to eliminate result difference caused by individual capability.

For the management and control system of production site of electrical system, any professional and standardized operation is represented by orderly management and control on site. Its basis and management and control objects mainly come from work safety and technical activities. For repeated work, according to the principle of perfect workers, different steps in the whole process are divided into action details and corresponding indicators and elements. These elements are listed according to the objective law that they must obey during the implementation process. These elements are arranged in a certain sequence of behavior and thinking logic and are combined into more reasonable operation procedures and steps. Besides, a standardized template in file format will be distributed to each implementation unit. Before implementation of work, the construction personnel shall combined the standardized template with the actual situation on site and prepare the final standardized document, and report it to the enterprise for approval. During the implementation of maintenance work, the task of construction personnel is to carry out construction in strict accordance with the approved standardized operation document without any reservation.

It should be noted that, standardized operation eliminates the operation result caused by difference in the capability of workers. In the meanwhile, it can prevent safety accidents. Such operation is very similar to list revolution advocated by different sectors. This chapter mainly focuses on the maintenance of power transformation equipment.

Module 1 Identification of On-site Hazards

The maintenance of electrical equipment involves various types and models of equipment, various test items, primary and secondary equipment, varying work site, and joint construction by external workers. Besides, most substation maintenance should be completed by people. Due to the difference in individual skills, experience, and cognitive level, even the same work is carried out by different workers, there may be difference in the identification of hazards on the work site. As a result, unsafe situation may occur during work. This section mainly explains the basic principles and key points of hazard identification and gives some cases, in order to enable readers to have a full understanding of some typical work hazards in substation maintenance.

1. Key Points of Hazard Analysis

During the implementation of substation maintenance, due to a certain difference in the specific working environment and maintenance conditions, the hazards and specific key points for analysis may vary from case to case. If the maintainer conducts maintenance at heights, the position of hazards will be very high, and the maintainer may fall from heights easily. For hot-line maintenance, the maintainer may also be exposed to obvious live parts. In addition, in many cases, there exist harmful gases in the substation maintenance environment. Such harmful gases, i.e. hazards, may cause injury to the maintainer and damage to the equipment under maintenance. In terms of substation maintenance, the equipment used by the maintainer is also a kind of hazard. For example, some large mechanical equipment like crane may become a hazard. If the maintainer makes any mistake during operation, such maloperation may cause injury to the maintainer. This is also a kind of hazard. Therefore, the maintainer should conduct operation in strict accordance with relevant procedures and specifications, in order to prevent safety accidents. In terms of the key points of analysis of hazards in substation maintenance, the maintainer should also recognize the significance of psychological safety, ease negative psychology, and overcome obvious fear of difficulty. The maintainer should ensure full energy and positive psychology in practical work. If they are in a poor psychological state, they will not have enough resistance to pressure in practical work, and even cause safety accidents due to distraction. Refer to Schedule 1.

2. Main Steps for Hazard Analysis

1) Sum up Experience and Refer to Relevant Data

Before the analysis of hazards in substation maintenance, relevant personnel should first take the nature, characteristics and main impact of hazards into full consideration, and predict potential accidents. Relevant personnel shall analyze the accidents and problems in previous substation maintenance, draw lessons from the past, and absorb excellent experience, in order to provide primary sources for substation maintenance. Besides, the maintainer shall combine the data of hazards in the past with existing maintenance environment for analysis, and use them for reference.

2) Clarify Responsibilities and Keep Communication

Hold a meeting to clarify the division of labor in advance. For potential hazards in maintenance, the maintainer should analyze them together with other workers, carry out intensive discussions, and jointly propose preventive measures. Ensure that a tester is assigned to be responsible for each corresponding position. Finally, sum up the data of each hazard and use them to guide work. Prepare reasonable safety precautions for identified hazards. Inform the operator of the actual situation without any reservation. No matter how accurate the evaluation is, the operator should conduct a field inspection during specific operation. This work is heavy and complex, and new conditions may occur at any time. The worker shall keep close communication with the protector, report problems to the superior once they are identified, and solve such problems in a proper and reasonable way.

3) Grouping Management and Emergency Plan Preparation

The maintainer should prepare an emergency plan according to the type of hazard. Besides, the responsibility of each maintainer and other worker shall be defined and different workers shall be divided into different groups, so as to facilitate emergency management.

3. Safety Control on the Work Site

1) Strengthen Process Management of Safety Work on Site

According to the current safety work process, a standardized work process shall be provided with corresponding standardized on-site safety measures. During each operation process, clarify the role and responsibility of each leader and relevant worker, and employ standardized process to manage each step which may threaten the equipment and personal safety. Take preventive measures to minimize the hazard. Eliminate hidden dangers at the very beginning. Identify and prevent any undesirable phenomenon in a timely and proper way.

2) Operation on the Maintenance Site

The basic procedures for the standard of safety measures on the maintenance site shall include site investigation, standard preparation, review by superior, standard implementation, summary, evaluation, and examination.

(1) Site investigation: In order to detect potential operation hazards in advance and prepare protective measures, the work unit shall arrive at the substation in advance and conduct site investigation. Cut off the power supply for equipment under test before taking safety measures, and install a grounding wire for live parts or close the grounding knife switch for the equipment under maintenance which may be electrified. Avoid waste of funds and unnecessary manpower, materials and financial resources caused by repeated safety measures. Meanwhile, take into account the practical situation of equipment maintenance, avoid maintenance and provide corresponding protective measures for the work site.

(2) Standard preparation: Prepare drawings and cards according to the results of site

investigation. Generally, the details of safety measures and positions of potential hazards required on the work site shall be filled in the section of unified measures. The actual operating condition of equipment and layout of safety facilities on site shall be indicated in detail on the layout plan of substation.

(3) Review by superior: The superior shall conduct strict and regular review of the cards prepared, in order to ensure implementation of safety measures for maintenance on site. The review shall be binding and enforceable.

(4) Standard implementation: Implement control measures for safety and hazard control on site should in strict accordance with the approved document, and clarify the responsibilities of relevant personnel, in order to ensure clear division of labor in case of emergency.

(5) Summary, evaluation and assessment: Summarize the work immediately after the maintenance is completed, evaluate the effect of the standard for safety measures on the maintenance site, draw lessons from the work, and improve the safety measures and procedures.

3) Ensure Monitoring

The power company and substation maintainers shall strictly monitor relevant equipment, which is running, have a clear understanding of the current operating condition and safety performance of equipment, assist substation maintainers in adjusting the performance of equipment, so as to ensure no hazard on the equipment. Obsolete old equipment timely, so as to prevent safety accidents during the period of operation or maintenance, especially for some non-electrical components which may cause personal injury once they are out of order, including energy storage spring, density relay, sealing device, gas or oil valve, steel wire rope for lifting. The status of these equipment and articles can hardly be displayed on the electricity protection device. Therefore, operation and maintenance monitoring shall be strengthened to provide a basis for determination of hazards in maintenance.

模块二　现场危险点工作案例

一、高压开关柜危险点分析及预控措施

1. 110 kV××变电站

全站 10 kV 开关柜柜型：GG-1A(F)-28G；

生产单位：四川开关厂；

出厂时间：2011 年 10 月。

表 2-1　GG-1A(F)-28G 高压开关柜危险分析及预控措施

序号	危险点	控制措施
1	母排未加装绝缘护套、运行时间较长后易积污（见图 2-1）	进行母线停电例检工作时，应加强对母线的清洁工作
2	10 kV 裸露铝排带电时，在开关间隔内搬运长物时，可能误碰触电（见图 2-2）	加强验收资料管理，例检和巡视到位，缺陷及事前上报
3	10 kV 裸露铝排带电时，在开关间隔内搬运长物，可能误碰触电 10 kV 开关柜总路进线与母线	1. 开关间隔内搬运长物，应两人放倒搬运。 2. 制订停电检修计划时，尽量采取分段或全站集中检修，使工作区裸母线桥不带电
4	10 kV 开关柜总路径线与母线间距离过近，当总路线和检修（尤其是主变侧刀闸的检修）而母线部停电时或母线停电检修而主变不停电时，存在易触碰带电体的风险	1. 在总路柜和隔离柜之间加装隔离挡板图。 2. 不满足作业安全条件时，总路与母线同时停电检修
5	处理 10 kV 出线侧穿墙套管缺陷时，距离母线较近，易发生触电危险（见图 2-3）	1. 在开关柜顶上母线和出线之间加装永久性隔离挡板。 2. 不满足作业安全条件时，母线与出线同时停电检修
6	10 kV 部分开关柜无法正常关合	有误碰带电设备的风险，操作中应加强监护
7	10 kV I 段母线，PT 柜无带电显示装置	在该设备进行操作时应加强验电工作
8	部分 10 kV 开关柜上沿无法关合	有误碰及小动物窜入的风险

图 2-1　母排未加装绝缘护套

图 2-2 穿墙套管与带电母线距离过近

图 2-3 35 kV 母线为裸露铝母线

2. 35 kV××变电站

全站 10 kV 开关柜柜型：KYN28-12；
生产单位：四川华仪开关厂；
出厂时间：2012 年 08 月。

表 2-2 KYN28-12 高压开关柜危险分析及预控措施

序号	危险点	控制措施
1	10 kV 高压室内过桥离地过近，有误碰的风险（见图 4-4）	进入高压室工作的人员应相互加强监护
2	全站设备无带电显示器	停电工作时应应加强验电工作
3	10 kV 高压室空间狭窄（见图 2-5）	在检修或者操作过程中应防止误碰其它设备
4	开关分合闸无挡板易引起误碰分合（见图 2-6）	考虑加装挡板
5	部分开关柜无法正常锁闭	停电检修工作中一并处理

 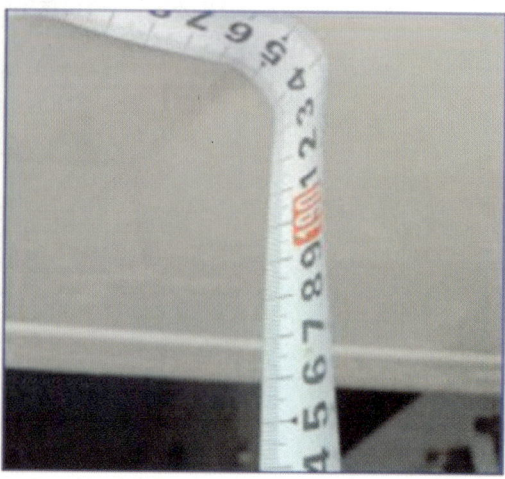

图 2-4　高压室内高桥高度太低不到 1.9 m

 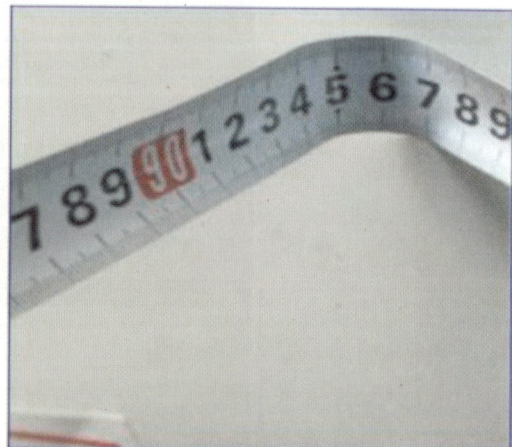

图 2-5　高压室内空间狭窄仅有 95 cm

图 2-6　开关柜开关面板无挡板易误碰

二、主变工作危险点分析及预控措施

220 kV××变电站

工作内容：Ⅰ#主变安装

表 2-3　220 kVⅠ#主变安装危险分析及预控措施

序号	作业内容	危险点	控制措施
1	搭接施工电源	防止触电	工作时设专人监护
			在站用变控制屏内设置临时绝缘隔板
		小动物进入低压室	封堵低压室门窗
	低压短路		在使用前应检查电源线是否合格
			应正确搭接电源
2	大型施工器材在施工现场的搬运	吊车在高压设备区行驶时误触带电体	吊车在进入高压设备区前，工作负责人会同吊车司机勘查和确定调车的行驶路线以及吊车与带电体的安全距离，明确带电部位、工作地点和安全注意事项
			吊车在高压设备区行驶时，必须设专人监护和引导
			工作前，工作负责人应向调车司机讲明作业现场周围邻近的带电部位，确定吊臂和重物的活动范围及回转方向。起吊作业必须得到指挥人的许可，并确保与带电体的安全距离准确数据。查阅《国家电网公司电力安全工作规程（火电厂动力部分）》第 16 部分"起重与运输内容"中表 16-2 和 16-3
			由专人指挥，指挥方式须明确且准确
			工作人员不得站在吊臂和重物的下面及重物移动的前方
			钢丝绳荷种需保证其安全系数。根据吊车吊臂角度确定荷载，不得超载使用
3	安装变压器引线、更换 110 kV 跨线	拆装引线时碰伤	工作时应戴手套和正确选用工具
		引线脱落、动荡、碰伤作业人员	引线拆装时，用传递绳或绝缘杆固定和传递
		高空坠落摔伤	高处作业人员必须使用安全带，穿防滑性能好的软底鞋
		个人伤害	现场工作人员必须戴好安全帽
			起重工器具使用前需按规定进行检查
			滑轮的悬挂点须牢固，滑轮门闭锁牢靠，钢丝绳套与滑轮间须采取防分离措施
		工具掉落	高处作业人员必须使用工具袋

续表

序号	作业内容	危险点	控制措施
3	安装变压器引线、更换110 kV跨线	一次水平引线放线过程中或落地后触及带电体	作业时由专人统一指挥并加设专责监护人
			用绝缘杆或绝缘绳固定水平引线与刀闸间的引线距离，确保其对带电体距离不小于下列数值，35 kV，1.0 m；110 kV，1.5 m；220 kV，3 m
		感应电压击伤，引发其他伤害	作业人员必须戴绝缘手套，系好安全带
			在有感应电压的场所工作时，应在工作地点加设临时接地线
4	附件安装	作业人员从器身顶部掉下	在器身顶部作业的人员必须使用安全带
			器身顶部的油污需清擦干净
		吊装时，附件脱落、摆动、放置不稳固、挤、碰伤作业人员	按标准选用的检查和起重工器具
			由专人统一指挥，明确起重工作的指挥方式
			控制吊装速度，保持重物平稳
			重物起调和下落过程中，作业人员未经工作负责人许可，不得擅自从事工作
		吊装套管时损伤套管	由专人统一指挥，明确协调工作的指挥方式
			由专人捆绑并竖立在专用设备上
			安装时应紧固螺栓后方可松绑
5	排油、注油和滤油	上下油缸和设备时，使用梯子不当而摔伤	梯子需放置稳固，由专人扶持或专梯专用，梯子与器身等固定物牢固地捆绑在一起
			上下梯子和设备上须清除鞋底的油污
		渗、漏油污染地面，滑倒、摔伤作业人员	设专人看管滤油设备，滤油点用容器盛接
			油管接头连接良好，油路密封良好
			作业人员须穿耐油性能好的防滑鞋
		发生火灾	作业现场严禁吸烟和明火，必须使用明火时应办理动火手续，并在现场备足消防器材
			作业现场不得存放易燃易爆品

Module 2　Cases of On-site Hazards

1. Analysis of Hazards of High-voltage Switch Cabinet and Preventive and Control Measures

1) 110 kV××Substation

Model of 10 kV switch cabinet for the whole substation: GG-1A(F)-28G

Manufacturer: Sichuan Switchgear Plant

Date of production: October 2011

Table 2-1 Analysis of Hazards of GG-1A(F)-28G High-voltage Switch Cabinet and Preventive and Control Measures

S/N	Hazards	Control measures
1	The busbar is not provided with an insulating sheath and dirt may accumulate after operation for a long period (As shown in Fig. 2-1).	Strengthen bus cleaning during the routine interruption maintenance of bus.
2	When the 10kV bare aluminum busbar is live, the operator handling long articles in the switch room may accidentally touch the busbar and get an electric shock (see Fig. 2-2).	Enhancement of acceptance data management, routine inspections and walk-throughs in place, defects and prior reporting.
3	When the 10 kV bare aluminium busbar is electrified, the operator who is handling long articles in the switch room may accidentally touch the busbar and get an electric shock.	1. For the handling of long articles in the switch room, two people shall work together to put articles down before handling; 2. As for the preparation of interruption maintenance plan, maintenance shall be carried out in sections or in the whole substation, and ensure that the bare bus bridge in the work area is not electrified.
4	The general incoming line of 10 kV switch cabinet is too close to the bus. When the general line is under interval maintenance (especially the maintenance of knife switch on the main transformer) but the bus is still electrified, or when the bus is under interruption maintenance but the main transformer is still electrified, there is a risk that the maintainer may touch charged objects.	1. An isolating partition shall be installed between the general line cabinet and the isolation cabinet. 2. If the requirements for operation safety are not met, interruption maintenance shall be carried out for the general line and the bus simultaneously.
5	During the handling of defect in the wall-through bushing on 10 kV outgoing line, the wall-through bushing is too close to the bus, which may easily cause electric shock hazard (As shown in Fig. 2-2).	1. A permanent isolating partition shall be installed between the bus and the outgoing line on top of the switch cabinet. 2. If the requirements for operation safety are not met, interruption maintenance shall be carried out for the bus and the outgoing line simultaneously.
6	Some 10 kV switch cabinets cannot be closed normally.	There is a risk of accidental touch with live equipment and monitoring shall be strengthened during operation.
7	PT cabinet of section-I 10 kV bus is not provided with an electricity display device.	Verification of live parts shall be strengthened during the operation of such equipment.
8	The upper edge of some 10 kV switch cabinets cannot be closed.	There is a risk of accidental touch and intrusion of small animals.

Fig. 2-1 Busbar without Insulating Sheath

Fig. 2-2 Too Short Distance between the Wall-through Bushing and the Live Bus

Fig. 2-3 The 35 kV Bus Is Bare Aluminium Bus

2) **35 kV××Substation**

Model of 10 kV switch cabinet for the whole substation: KYN28-12

Manufacturer: Sichuan Huayi Electric Company Limited

Date of production: August 2012

Table 2-2　Analysis of Hazards of KYN28-12 High-voltage Switch Cabinet and Preventive and Control Measures

S/N	Hazards	Control measures
1	The bridge in 10 kV high-voltage room is too close to ground and there is a risk of accidental touch (as shown in Fig. 2-4).	Workers who are working in the high-voltage room shall strengthen monitoring of each other.
2	Equipment at the whole substation is not provided with a live display.	Verification of live parts shall be strengthened during interruption.
3	The 10 kV high-voltage room is too narrow (as shown in Fig. 2-5).	The operator shall prevent accidental touch with other equipment during maintenance or operation.
4	There is no baffle for switch opening and closing, which may cause accidental opening or closing (as shown in Fig. 2-6).	A baffle shall be installed.
5	Some switch cabinets cannot be closed normally.	Such problems shall be handled together during interruption maintenance.

 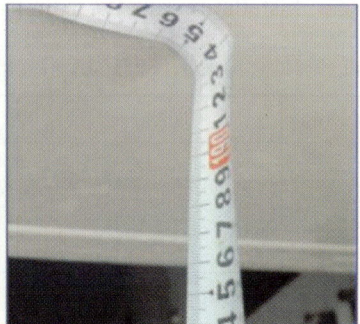

Fig. 2-4　The Bridge in the High-voltage Room Is Less Than 1.9m from the Ground

Fig. 2-5　The High-voltage Room Is Too Narrow (Only 95cm)

Fig. 2-6　There Is No Baffle on the Switch Cabinet Panel, Which May Cause Accidental Touch

2. Analysis of Hazards of the Main Transformer and Preventive and Control Measures

220 kV × × Substation

Work content: Installation of I # main transformer

Table 2-3　Installation Risk Analysis of 220 kV I # Main Transformer and Preventive and Control Measures

S/N	Scope of work	Hazards	Control measures
1	Joining the construction power supply in an overlapping way	Protection against electric shock	Specially assigned a person for supervising.
			Install temporary insulation partitions in the control panel of the station transformer.
		Small animals entering the LV room	Seal the doors and windows of the LV room.
		LV short circuit	Check if the power cord is acceptable before use.
			Properly join the power supply in an overlapping way.
			Before the crane enters the high-voltage equipment area, the person in charge of work shall investigate and determine the route of the crane, check the safe distance between the crane and the charged object, and specify the live parts, work site, and safety precautions.

Continued

S/N	Scope of work	Hazards	Control measures
2	Handling of large construction equipment on the construction site	Unintended touch of an charged object when the crane moves in the high-voltage equipment area.	If the crane moves in the high-voltage equipment area, a special person must be appointed to supervise and guide it.
			Prior to the work, the person in charge of work shall inform the crane operator of the live parts around the construction site and determine the motion range and rotation direction of the jib and heavy weights.
			Lifting operations must be approved by the commander, and a safe distance from the charged object must be guaranteed. See Tables 16-2 and 16-3 of Part 16—Lifting and Transportation of the *Electric Power Safety Working Regulations (Power of Thermal Power Plants) of State Grid Corporation of China* for specific data.
			A commander shall be specially assigned to command the lifting operations, and the commander shall command clearly and accurately.
			Workers shall not stand under the jib and heavy weight or in front of the moving heavy weight.
			The lifting and rotation speed shall be controlled to ensure smooth lifting and moving.
		Crashes and collisions of equipment during lifting and placing	A person in charge shall be specially assigned to command the lifting operations, and the person shall command clearly and accurately.
			Workers shall not stand under the jib and heavy weight or in front of the moving heavy weight.
			The load of the wire rope shall be adequate to ensure its safety factor. The load of the crane shall be determined according to the angle of the jib, and the crane shall not be used in the event of overload.

Continued

S/N	Scope of work	Hazards	Control measures
3	Installation of transformer leads and replacement of 110 kV jumper wires	Hitting injuries caused during installing and removing leads.	Wear gloves during operation and select correct tools.
		Operators injured by any falling and swinging leads.	When installing and removing leads, use a hauling rope or insulating rod to fix or haul the leads.
		Injuries caused by falling from heights.	The personnel working at heights must wear safety belts and soft-soled shoes with good antiskid performance.
		Personal injuries.	On-site operators must wear safety helmets.
			The lifting tools and instruments hall be checked as required before use.
			The suspension point of the pulley shall be secured, the pulley door shall be locked reliably, and measures shall be taken between the wire rope sleeve and the pulley to prevent them from separating.
		Falling of tools.	Operators working at heights must use tool bags.
		A primary horizontal lead touching a charged object during paying off or after falling to the ground.	The operations shall be commanded by a specially assigned commander and supervised by a specially assigned supervisor.
			The horizontal lead and the knife switch lead shall be fixed with an insulating rod or insulating rope to keep a fixed distance between them, and make sure that their distance from the charged object shall not be less than the following values: 35 kV, 1.0 m; 110 kV, 1.5 m; 220 kV, 3.0 m.
		Other damage caused by induced voltage injuries.	Operators must wear gloves and safety belts.
			When working in the place where the induced voltage exists, operators shall install more temporary grounding wires on the work site.

Continued

S/N	Scope of work	Hazards	Control measures
4	Installation of accessories	Operators falling from the top of the transformer body.	Operators working on the top of the transformer body must wear safety belts.
			The oil stains on the top of the transformer body shall be cleaned.
		Operators squeezed by and collided with any accessories that fall off, swing, or are placed unstably during lifting.	Tools and instruments shall be chosen and checked according to standards.
			A commander shall be specially assigned to command the lifting operations, and the way of commanding shall be specified.
			The lifting speed shall be controlled to keep the heavy weight stable.
			During the lifting and dropping, any operators shall not work without the permission of the person in charge of work.
		Bushings damaged during lifting.	A commander shall be specially assigned to command the lifting operations, and the way of commanding shall be specified.
			The bushing shall be bound by a specially assigned person and set upright on special equipment.
			When installing the bushing, the bolts shall be tightened before unbinding the bushing.
5	Oil discharge, injection, and filtration	Injuries caused by falling from any improperly used ladders when ascending and descending the cylinder and equipment.	The ladder must be firmly placed and supported by a specially assigned person, or a special ladder shall be used, and the ladder and other fixed objects such as the transformer body must be firmly tied together.
			Oil stains shall be removed from the soles when ascending and descending the ladder and equipment.
		Operators injured from sliding and falling down to the ground contaminated by any oil leaks and spills.	A person shall be specially assigned to attend to the oil filtering equipment, and a container shall be used to catch any leaking oil at the leaking point.
			The oil pipe shall be well connected and the oil line shall be well sealed.
			Operators shall wear antiskid shoes with good oil resistance.

Continued

S/N	Scope of work	Hazards	Control measures
5	Oil discharge, injection, and filtration	Outbreak of fire.	Smoking and, open flames are strictly prohibited at the work site. If it's necessary to use open flames, a hot work permit shall be applied. Formalities shall be handled and adequate fire equipment shall be provided on site.
			Flammable and explosive materials shall not be stored on site.

模块三 标准化作业

通常来讲，变电检修现场标准化流程包含了三个阶段，即现场作业准备、检修作业实施、检修作业总结完结。每个阶段又可以分为多个步骤。

一、变电检修现场作业准备

（一）获取工作任务

1. 工作计划的下达途径

通常情况下，在国网系统中，变电检修计划大致可以分为三个种类，即计划检修、一般缺陷消除和危重故障处理。

计划检修（下称检修计划）的制订是逐级完成的。上级设备主管部门（以下简称设备部）是变电检修计划制定主体。检修计划的制订时间一般始于每年9月，并于当年的12月完成下发。检修计划编制素材的来源包含：检修、运维、物资、规划和用户等多个与系统和设备运行有关的方面。在国家电网下属的地市级别供电企业中，所辖设备及供电营业区客户设备的停电检修工作任务是由下面的部门计划、协调、安排的。

上级设备主管部门（简称设备部，下同）是该公司所属电网输变电设备及城市配网检修工作的归口管理部门，负责组织、协调、制订年度检修计划，审批检修项目，检查检修计划执行情况，考核检修工作的安全、进度和质量，并组织对重大检修项目的验收。

公司调度控制中心（简称调控中心，下同）负责组织审核各运行单位上报的检修停电计划。负责制定网内输变电设备停电检修计划的总体方案，确定检修期间电网运行方式安排，签发、执行调度检修申请书，监督输变电设备检修计划执行情况，统计各单位检修工作完成情况，配合设备部制订检修项目安排和对输变电设备检修情况进行考核。

公司营销（农电）部负责协助并督促职责范围内的客户制订检修计划，审核需对直供客户停电的检修工作计划，并纳入电网年度、月度停电检修计划。营销（农电）部同时也是该供电公司所属农村配电网络及设备检修工作的归口管理部门，负责组织、协调、制订年度检修计划，审批检修项目，检查检修计划执行情况，考核检修工作的安全、进度和质量，并组织对重大检修项目的验收。

2. 检修计划的下达流程

设备部将设备检修计划上报省公司设备部，省公司设备部汇总全省情况后，加入 220 kV 以上电压等级的检修需求并上报国网西南分部等区域电网公司，依此类推，逐层增加后，直到国网公司。国网公司协调各区域电网公司的检修计划，追加并指定部分需要跨区域或重点计划的时间节点，以此形成检修计划的预安排。下发给各区域电网公司，同样指定部分工作的时间节点后下发省公司，然后下发至地市公司设备部。最后下发到基层检修单位。基层检修单位运检管理人员按照规定的时间节点准备必要物资的采购与调用，并将其余未规定时间的任务拆解为季度计划和月度计划，并下发至班组。班组长将月度计划最终落实为周计划，并指派工作负责人实施工作。具体流程如下：

1）省级调度管辖设备的停电计划下达流程

（1）年度计划：由运检、建设等业务主管部门、检修管理单位、施工单位负责编制下一年度省调管辖设备的停电需求，经设备部审核，分管领导批准后，由调控主管部门转报省调，经省调平衡后下达执行。

（2）季度计划：由设备部、建设部、检修管理单位、施工单位负责按年度停电检修计划编制下一季度省调管辖设备的停电需求，经设备部审核，分管领导批准后，由调控中心转报省调，经省调平衡后下达执行。

（3）月度计划：由设备部、建设部、检修管理单位、施工单位负责按年度停电检修计划编制次月省调管辖设备的停电需求，经设备部审核，分管领导批准后，由调控中心转报省调，经省调平衡后下达执行。

（4）周计划：检修施工单位根据省调季度及月度检修计划，于开工前一周将下一周的检修计划报送至调控中心，再由调控中心转报省调。

2）地市级供电公司调控中心管辖设备的停电计划编制流程

（1）年度计划：由运检、建设、营销等业务主管部门、检修管理单位、施工单位将各自管辖范围的下一年度停电需求报送至调控中心汇总。营销部收集 35 kV 及以上大宗用户下一年的停电需求，汇总后报送至调控中心。调控中心会同有关部门编制"电网设备年度检修计划"，报供电公司批准。

（2）月度计划：各检修、施工单位每月向设备运行单位提供次月停电需求，设备运行单位在收到停电需求后，不得删减检修、施工单位的停电需求，不同的意见在公司月度检修平衡会上提出。设备运行单位经内部初步平衡后，同时向调控中心、运检、营销等业务主管部门报送次月停电需求计划。营销主管部门收集各专线用户次月的停电需求，汇总后报送至调控中心。调控中心会同有关部门综合平衡后，召开由公司分管领导参与的月度检修平衡会。

（3）各县级分公司于每月市公司月度检修平衡会后将各自管辖范围内的月度停电检修计划正式版（分管领导签字盖章）报送至市公司调控中心。由市公司调控中心汇总编制"电网设备月度停电检修计划"，报市公司批准下达。

报送的停电需求计划应当详细、清楚，要求停电的设备必须写明具体停电范围、主要工作内容、停电时间和其他特殊要求等。

3. 检修计划的下达形式

检修计划的管理分为：计划检修和非计划检修。

计划检修是指纳入年度、月度、周安排的停电计划，且按调度部门批复的时间进行的设备检修、维护、试验、改造、基建等停电工作。

非计划检修是指除计划检修外的所有检修，分为一般缺陷消除和危重故障处理。

（1）一般缺陷消除是指在计划外，日常巡查或运行中额外发现的一般缺陷，通常由运维或检修部门上报需求，并在下一次停电检修计划中一并处理。通常而言，这类检修任务都会有一定的准备时间。

（2）危重故障处理是指设备在运行中突发的紧急故障，如不及时消除故障会严重威胁系统正常运行。通常来说，此类任务要求检修人员在得到运维部门的检修需求报告后立即赶赴现场进行处理。

（3）理论上每个检修人员都可以在本单位运检管理人员那里获知本年度的检修计划，在班长处可以得到本月的计划内检修任务，使得检修人员具备了提前准备的可能。相对于没有准备时间、更难以针对练习的危重故障处理任务，计划检修、一般缺陷消除任务留有足够的时间给检修人员做针对性准备，就好比开卷考试。对于检修人员而言，开卷考试是自身检修业务积累和提高的过程，故障处理是对人员检修水平的考验。

（二）实施现场查勘

1. 现场查勘的依据

根据《电力安全工作规程（变电部分）》（以下简称《安规（变电部分）》）的规定，变电检修（施工）作业，工作票签发人或工作负责人认为有必要现场勘察的，检修（施工）单位应根据工作任务组织现场勘察，并填写现场勘察记录。现场勘察由工作票签发人或工作负责人组织。

虽然规程中对现场进行查勘的时机是在"有必要"时，但由于无人值守变电站的规模不断增大，建议对于常规的A、B、C、D类检修或作业场地周边涉及带电部位的时候，为确切了解最新的实际情况以便于更好地做出准备，都应进行查勘。

2. 查勘的形式

必须是人员到现场实地查勘。时间建议在计划开工前7~15天。这个时间不宜太靠近或太提前。太靠近会导致开工的其他准备流程时间不足，并且需要停电的工作至少要提前一个星期办理停电申请，为调度部门预留足够的负荷转移及下发停电通知的时间。以7天为例计算：停电申请的审批至少需等待2天，一种工作票至少需提前1天，剩余的4天需要完成方案编制、审核、物料准备和人员学习，其实已经非常紧。太提前也不可取，因为可能会导致查勘信息在时效性上出现偏差。

3. 查勘的内容

在《安规（变电部分）》中，并没有明确规定查勘的内容。这里引用《安规（变电部分）》线路部分中的5.2.2条款：现场勘查应查看现场施工（检修）作业需要停电的范围、保留的带电部位和作业现场的条件、环境及其他危险点等。实际工作中，需要查勘的内容需要注意的细节其实很多，可参考附表2。

（三）施工方案的编制

根据现场勘察结果，对危险性、复杂性和困难程度较大的作业项目，应编制组织措施、技术措施、安全措施，经本单位分管生产的领导（总工程师）批准后执行。

施工方案的内容：施工方案是施工准备内容的核心，是检修工作实施的头脑预演和准备依据，还是本次工作中人员分工、检修内容和技术标准的书面依据，更是施工流程和进度安排的逻辑梳理。通常而言，施工方案是由工作负责人执笔，以文书的方式进行编制，尽管格式并未统一，但在内容上，大致都可归类为组织措施、技术措施、安全措施三个方面，所以检修方案也被称为"三措"。组织措施主要用于明确与"人"相关的资源配置，技术措施全部用于明确与"物"相关的的资源配置，安全措施则旨在针对上述两点做出安全修正。

三者相互独立但又相互关联，在逻辑上互相补充、观点上相互针对，这一点在这三者的编制过程中会始终贯穿，共同保证方案的完整、正确、客观。

组织措施主要需要解决以下4个问题：任务需要设置的人员分工岗位有哪些？每个分工岗位的技术和安全职责是什么？每个分工岗位需要的人数和能力要求？以上岗位最佳的人选有谁？

技术措施主要需要解决以下问题：检修任务具体内容如何排序？技术和质量控制标准是什么？技术和质量控制需要哪些物料作为支撑？现场实施的时间进度安排是什么？施工现场如何布置施工，实际成本估算是否满足预算？

（四）标准化作业文件的审批

现场作业使用的各类标准化作业文件，必须履行相应的审批手续，实行责任跟踪制度。审批时必须全面审核作业的全过程，重点审查危险点分析、控制措施、工序等内容。审批完毕后，审批人员必须签署具体意见（含修改意见）。

1. 审批原则：按作业管理组织关系履行标准卡分层审批

（1）110 kV 及以下输电线路和辅助设备大修、220 kV 及以下输变电设备 C、D 类常规检修与巡视的标准卡由专业车间（室）相关技术人员审核，主管（副主管）批准。

（2）35 kV 及 110 kV 主变大修、220 kV 辅助设备大修、220 kV 输电线路大修的施工组织方案或标准卡由专业车间（室）和运维部专工审核，报设备部（副）主任批准。

（3）35 kV 及 110 kV 设备的技术改造工作的施工组织方案或标准卡由专业车间（室）和设备部专工审核，报设备部（副）主任批准。

（4）220 kV 主变大修、220 kV 主设备技术改造工作或经设备部确定的其他特殊和大型施工任务的施工组织方案或标准卡由专业车间（室）和设备部审核，安全监察部会审，报供电公司生产副总（总工）批准。

2. 标准化作业文件的学习

所有经审批后的作业文件，应当在作业之前由班长组织召开班前会，进行逐一学习。

二、变电检修现场作业实施

在完成检修作业的各项准备工作后，按照批复的检修时间，工作负责人应携带提前准备的标准化工序卡、工作票、安全学习记录带领工作班成员到达工作地点。出发前，工作负责人应再次确认天气状况、物料和工作人员的到位情况。

一般非特殊情况下，调度在批复检修时间时，为了保障充足的检修时长和避开夜间施工带来的不利和安全隐患，检修开始时间通常都会被安排在上午。运维人员对检修设备的停电倒闸操作和安全措施的设置则通常被安排在前一天的后半夜进行，当前由于运检一体化的推行，该停电工作的人员和时间可能各单位不同。如果停电操作受到拖延，检修人员到场后，运维人员未能完成安全措施设置，则检修人员会被要求原地等待。此时，检修人员理论上不应在变电站逗留，但在实际工作中并未如此严格，但检修人员应集中不分散，绝对不得进入变电站设备场地或中控室。待运维人员完成安全措施后，许可人联系工作负责人办理工作许可手续。

（一）执行现场工作许可手续

执行现场工作许可手续，工作许可人会同工作负责人应到现场确认工作票上所列的安全措施已设置无误，工作许可人应向工作负责人交待清楚停电范围并当面证明设备确实停电。双方无疑问后在工作票上分别亲自签名，工作负责人和许可人各持一份工作票，回到各自岗位。工作负责人持票带领工作班成员进入场地实施开工前的准备工作。工作许可人持票留存并将许可开始工作时间填写在运行记录簿内。

需特别指明的是工作班成员只有在完成现场工作许可手续后，并且在工作负责人的带领下才能进入现场。

（二）召开班前会

由于执行现场工作许可手续的只是工作负责人本人，工作班成员并不参加。为此，工作负责人必须召开班前会进行安全学习，内容是所有经审批后的作业文件内容。需要工作负责人告知并确保每一个工作班成员了解，这也被称为现场安全技术措施交底程序。班前会的形式比较统一，多为工作班人员空手现场列队，工作负责人首先检查成员的着装及精神状态，询问工作班成员有无异常或要求。然后工作负责人进行宣讲，宣讲至少要有"四清楚"，即工作任务作业流程清楚，人员分工清楚，工作现场的安全措施、危险点及控制措施清楚，工作现场的停电范围、保留带电部分清楚。在宣讲过程中，工作班成员可以提问。工作负责人宣讲完毕后应进行现场抽问，确保所有工作班人员均已掌握无误。

工作班组人员在确认工作负责人布置的任务和本施工项目安全措施后，均应履行签字确认手续，在工作负责人收执的工作票上亲笔签名。签字后的工作班成员方可参加工作。

采用分工作票时，分工作负责人在总工作票上签名，其余工作班人员在分工作票上签名。

（三）实施现场作业

现场作业的实施阶段，其实就是对检修方案的落实阶段。检修人员按照组织措施的分工，履行技术措施的工艺和标准，同时严格按照安全措施的要求规避危险。工作负责人的主要任务是指挥、监护、确认工序和质量，工作班成员则是实施具体操作。

那么如何保证细节繁多的技术方案能够在现场得到有序开展呢？这就需要依靠标准化工序卡发挥作用了。标准化工序卡可以视作检修方案的具体执行版本，与工作票一样需要工作负责人在工作中随身携带。标准化工序卡的作用在于提醒和参照。因为没有任何流程或文件要求，检修人员在检修中必须依靠记忆进行施工。通俗地讲，任何一次列入工作计划的检修任务都有至少一个星期的准备时间，这对于检修人员而言其实类似开卷考试。只有在紧急事故或缺陷的处理上，由于时间的原因导致无法在事前详细编制检修方案，才需要在现场考验检修人员的技术积累。

因此，工作负责人和工作班成员在作业中应当充分利用标准化工序卡的提醒和参照作用，将自己的精力和注意力从现场流程与标准提取上转移到操作的实施与管控上。简单地说，就是将事前就能固化的正确内容写在纸上，到了现场依葫芦画瓢就行，人员在现场需要关注的是画得像不像、画得安不安全。

标准化工序卡的构成是按照工序组成的，由技术标准中的操作项目组成操作步骤。每个操作步骤在实施中都按照相同循环方式环状执行，循坏共分为四个环节：

下达：工作负责人按照标准化工序卡的项目顺序逐项下达指令并诵读技术标准与危险。

实施：工作班成员复令后严格按照技术标准中规定的方法与工艺进行作业操作。

核查：工作班成员完成操作后，工作负责人参照技术标准中的复核方案进行检查。

确认：工作负责人核对参数无误并做记录后，由操作人员在工序卡上的本步骤检修人员栏签名。工作负责人在工序卡的本步骤确认栏上打勾并签名，本步骤的工作结束，开始下一步骤循环的下达环节。如果该步骤产生了检测打印报告等记录原件，操作人和复核人同样需要在原件上签字并注明时间，如果该记录为电子档，则应当由工作负责人将文件名与保存地址在工序卡中注明，建议工作负责人使用独立的移动储存现场备份。

作业过程中，如发现与现场实际、相关图纸及有关规定不符等情况，工作负责人应根据现场实际情况及时修改工艺卡，必要时告知原审批人并经同意，方可继续作业。作业结束后，原审批人应履行补签字手续。

作业过程中，如发现设备存在事先未发现的缺陷或异常，作业人员应立即汇报工作负责人（小组负责人），并进行详细分析，制定处理意见。作业人员处理完毕后的缺陷和异常，在现场作业工序卡对应栏中如实、逐一填写，工作负责人（小组负责人）必须经确认验收后，在"确认人"对应栏中签字确认，严禁简单打"√"。

不能处理的缺陷和异常，在现场作业工序卡中进行如实、逐一填写，并在发现人中签字确认，工作负责人（小组负责人）在"确认人"处签字确认。

缺陷和异常导致设备不能正常投运时，工作总负责人应告知原审批人并经同意，方可继续作业。作业结束后，原审批人应履行补签字手续。

作业过程中，如遇到意外事件被迫变更作业流程、作业项目等情况，作业人员应立即汇报工作负责人（小组负责人）。工作负责人（小组负责人）应及时提出变更、调整意见，并在现场作业工序卡作业流程、作业项目变更情况栏中逐一填写，发现人和工作负责人（小组负责人）在"申请人"处逐条签字确认。必要时告知原审批人并经同意，重新调整作业内容后方可继续作业。作业结束后，原审批人应履行补签字手续。

在这个过程中，要求现场作业人员的行为应当完全参照事前编制的文本内容进行，原则上不需要也不能擅自做出变动。避免变动的根本原因并不是不愿意接受思考和创新，而是不接受没有经过事前慎密计划或设计的工作，避免出现没把握而导致的错误。所以"不能变动"的对象是指已经通过计划、校验、审核的方案、工序、工艺和标准，而范围也仅限于现场作业的实施环节。因此，如果在忽略或无视在检修方案编制中细致、严谨、完整、可操作的策划及准备，检修人员在实施操作中就必然面临没有依据或依据太过笼统无法实施的困境，对于检修工作本身而言这就是混乱的根源，在这种情况下人员安全和检修质量根本无从谈起。

三、变电检修现场作业完结

1. 填写现场检修记录

在检修测试中，按照状态检修三年三周期的纵比分析原则，检修人员应当将检修数据原始资料进行详尽记录并妥善保存，作为今后检修工作的依据和重要追溯信息。同时需要在变电站控制室内当值运维人员处填写修试记录，填写本次检修任务的内容、已处理的部分、仍遗留的问题，在记录末尾需要明确描述本次检修的结果：是否完成，能否投运。

2. 进行三级验收

在检修人员按照标准化作业工序卡的流程和内容完成各项操作并复核确认无误之后，需要将被检修设备的位置和设置恢复至检修前的初始状态，并对作业场地进行清理，工作班组自设的安全措施应由班组自行拆除。

作业场地清理的内容：废弃物不得随意排放，尤其不能排放在变电站排水沟中，以防污染站外的水道系统，必须按照规定进行回收和处理。对地面进行清洁，如有油污需要进行清洗。之后需要将工器具和物料转移到围栏出入口处，但不要装车，这是为了预防在接下来的 3 级验收流程中处理发现的问题。

输变电设备的年检预试及维护工作的质量验收实行 2 级验收制，大修、技术改造工作质量验收实行 3 级验收制。

申请验收的条件包括：

（1）检修项目和内容已全部完成。
（2）检修质量达到检修标准。
（3）检修记录、试验报告完备，签章手续齐全。

因此，在正式申请验收前，工作负责人会对本次检修进行班组自检，着重检查上述三点的完成情况。

第一级验收是指检修单位自检。检修单位专业技术主管人员作为本次检修管理方进行验收。由工作负责人向检修单位技术主管部门（设备部）提出申请，通常为运检技术主管或专责到现场进行验收。除了检修质量，还要关注检修任务的开展情况、设备遗留的问题等，目的是为下一步生产任务的安排提供支撑。

第二级验收是指运行单位对设备进行验收。由检修单位专业技术主管人员向运行单位技术主管部门提出验收申请。运行单位技术主管人员到场后与许可人和当天值班运维人员一道进行验收。不同的是，在这一级验收中，焦点由检修角度转换为使用角度，作为使用方，运行单位作为使用方将会更多地关注设备的遗留故障、远方就地旋钮操作、状态指示、信号反馈和整组传动等与运行状态相关的功能。

第三级验收是由供电公司组织进行验收。通常与第二级验收流程同时进行，由供电公司各相关专业技术主管人员到场后进行验收。

所有项目验收均应实行（现场）签字责任制和质量追溯制，验收中如发现不符合项，会出具不符合项通知单，并按相应程序处理。

3. 办理工作终结手续

完成验收后，检修人员就可以将本次携带的工器具搬运装车，车辆在装车完毕后应尽快回到指定位置等待。

工作负责人在其他非设备区域召集所有工作班成员召开班后会，与班前会相呼应，人员同样需要列队，同样由工作负责人主持，班后会的内容主要是确认无误之后，需要将被检修设备的位置和设置恢复至检修前的初始状态，并对作业场地进行清理，工作班组自设的安全措施应由班组自行拆除。

作业场地清理的内容：废弃物不得随意排放，尤其不能排放在变电站排水沟中以防对污染站外的水道系统，必须按照规定进行回收和处理。对地面进行清洁，如有油污需要进行清洗。之后需要将工器具和物料转移到围栏出入口处，但不要装车，这是为了预防在接下来的3级验收流程中处理发现的问题。

输变电设备的年检预试及维护工作的质量验收实行2级验收制，大修、技术改造工作质量验收实行3级验收制。

申请验收的条件包括：

双方签名后由工作许可人在工作负责人所收执的工作票上盖上"已执行"印章后，该工作视为终结。工作终结后，设备与场地视为带电，任何人不得再次进入工作地点，更不得擅自开展工作。许可人收执的工作票留运行继续办理工作票终结手续；工作负责人收执的工作票，需带回检修单位保存一年。

此时检修人员方可离开变电站。为了保障检修设备正常送电，变电检修和继电保护专业需要安排人员（通常为负责人）和工器具在站内留守至设备投运成功，在此期间留守人员的行为应当视为工作许可手续未办理时的要求。

四、现场检修完成后的总结和评价

对检修中的安全、质量、项目、工时、材料和备品配件、技术监督、费用等进行总结和评估。

施工单位应及时提交竣工报告、技术总结、试验报告、验收报告。

按照检修和技改项目管理规定要求整理检修档案，交公司档案室统一存档，同时交运行单位完整资料一份，检修单位档案保存周期为一个检修周期。

由于检修质量和验收不严造成事故,应分别追究检修人员和验收人员的责任。

分析设备异常状况及检修管理过程中的经验、教训的分析、总结,以不断提高检修质量和管理水平,提高设备的可靠性。

电网输变电设备年度检修任务完成后,各单位在每年 11 月对本年度检修计划完成情况进行统计和总结,经单位主管领导审核后报公司设备部和调控中心,同时各地市公司及以上级别的检修专责都能在 PMS 上调看年度完成情况、评价情况、闭环情况。

工作负责人的具体工作包括在 3 天内将收执的工作票交技术专责保存。在检修工作记录上进行登记,登记的内容包括工作的内容、开始和结束时间、地点、人员、发现的问题、处理的方法与结果。检修工作记录除了作为该工作团队现场工作量的凭证,还会用于班组今后查询设备检修记录的纸质留底。牵头编制检修报告,按照逐项确认后的标准化作业书将本次工作的过程数据中收集带回的检测结果和测试数据数据进行整理,按照指定的格式进行录入留存。对于在工作中产生的检测报告、测试数据资料原件,需要一并附在纸质报告中提交技术专责存档。这些资料同时也是星级班组的建设内容之一。

附表 1 变电检修典型设备工作危险点辨识及预控措施表通用部分

序号	工作项目	伤害类型	危险源	控制措施
一	现场查勘	直接:触电 间接:设备损坏	1. 查勘组织不力	1.1 大型停电检修工作的现场查勘,应由检修单位生产技术部门负责人组织,工作负责人、各专业小组负责人、现场安全负责人必须参加;其他停电检修工作的现场查勘,应由工作负责人组织,相关专业小组负责人参加。 1.2 查勘人员必须具备相应的安全知识和技能
			2. 误碰带电设备	2.1 明确查勘范围,与带电设备保持足够的安全距离。 2.2 严禁移动、翻越、开启围栏
			3. 查勘不到位	3.1 参加现场查勘的人员应事先了解清楚工作任务及具体工作项目,避免临时性增加工作内容。 3.2 认真查阅图纸资料,技术台账。 3.3 工作负责人及小组负责人必须核对所检修设备异动情况。 3.4 工作负责人及小组负责人核对上年度及本年度的缺陷记录和运行记录。 3.5 对工作现场的工作条件、工作范围的查勘必须到位,核实停电范围,必须明确保证作业安全的现场安全措施内容及布置要求。 3.6 查勘人员必须亲自到检修现场进行实地查勘,并做好查勘记录,查勘活动宜全程录音或录像。

续附表

序号	工作项目	伤害类型	危险源	控制措施
一	现场查勘	直接：触电 间接：设备损坏	3. 查勘不到位	3.7 强雷、暴风、暴雨等恶劣天气严禁进行现场查勘，查勘中出现恶劣气象时应立即停止查勘工作。 3.8 严禁单人进行查勘工作，严禁单人滞留在高压室和高压设备区。 3.9 查勘进入SF_6设备室、电容器室前必须按规定进行通风。进入电缆沟、电缆竖井等长期密闭环境前必须进行氧含量和有害气体检测。 3.10 查勘活动中严禁试图操作所有运用中的电气设备，严禁抽拉运行中的端子排接线，防止误碰裸露的低压缆线、接线端子和连接片，防止误碰导致二次空气开关跳闸
			4. 现场照明不足	查勘现场照度足够，必要时应使用移动照明设施
二	作业机具、安全工器具和材料准备	直接：机械伤害、火灾 间接：设备损坏	作业机具不合格、准备不到位、不按规定搬运	1.1 搬运较大或笨重器材时，不得直接用肩扛，应使用绳索和抬杠抬运。需要多人抬运的物件，须有专人指挥，统一信号，步调一致。 1.2 雨雪天运输器材时应注意防滑，在陡坡地段抬运时，路面上应采取防滑设施，同时要减轻每人所抬重量。 1.3 运输所用的抬运工具应牢固可靠，每次使用前，应由工作负责人进行检查，已经霉烂的绳索不得使用。 1.4 用跳板或圆木装卸滚动物件时，应用绳索控制物体，剩余物品要固定好，物件滚落前方严禁有人。 1.5 使用起重机装卸时，必须遵守起重机械安全管理规定。 1.6 作业机具工况良好，安全工器具合格，型号和数量满足工作要求。装箱前应事先检查确认，严禁带缺陷使用。 1.7 严格执行机具管理制度，做好定期检修、维护和保养工作。 1.8 易燃易爆材料、危险化学品、有毒物件的搬运与存放应符合安全规定
三	交通运输	直接：人身伤害、设备损坏	1. 无证驾驶车辆	1.1 严格驾驶员准驾车辆资格审查
			2. 病车上路	2.1 严禁车辆带病上路。 2.2 高温季节应重点检查轮胎温度、气压是否正常

续附表

序号	工作项目	伤害类型	危险源	控制措施
三	交通运输	直接：人身伤害、设备损坏	3. 车辆超载、超高、超宽、超速	3.1 超高、超宽车辆进入变电站应有专人引导。 3.2 装运设备、线盘、材料等容易滚（滑）动的物件，必须绑扎牢固，防止前后左右滚动。装运超长、超高物资，一律要有明显标志，按交通管理部门规定的时间、路线、速度行驶
			4. 防振动措施不当	采取适当的防振措施
			5. 疲劳驾驶、酒后驾驶、服用违禁药物后驾驶	严禁驾驶员疲劳驾驶、酒后驾驶和服用违禁药物后驾驶任何车辆
四	安全措施确认	间接：人身伤害	1. 工作票不合格	1.1 工作票签发人、工作负责人认真填写工作票。 1.2 工作票签发人和工作许可人要对工作票认真审核
			2. 现场安全措施不完善、不正确	工作负责人和工作许可人一起确认安全措施正确、完善后方可办理许可手续
五	工前交底	间接：人身伤害，设备损坏	1. 不进行工前交底	开工前必须进行工前交底
			2. 工前交底交代不清	工作负责人组织工作班成员认真学习工作票和进行安全技术交底，所有人员做到"四清楚"（作业任务清楚、危险点清楚、作业程序清楚、安全措施清楚）
			3. 工作班成员不清楚"四清楚"	3.1 随机进行"四清楚"抽查式询问。 3.2 工作班成员不清楚时要主动询问
六	工作间断、转移	直接：触电伤害	1. 未正确复核安全措施	复工时应得到工作负责人许可，取回工作票，工作负责人重新检查安全措施是否符合工作票的要求
			2. 误入间隔	2.1 若无工作负责人或专责监护人带领，工作人员不得进入工作地点。 2.2 工作负责人在转移工作地点时，应向工作人员交代电范围、安全措施和注意事项
七	工作终结	间接：设备损坏、电网事故、触电伤害	1. 临时安全措施未拆除	1.1 工作负责人必须认真检查临时安全措施已完全拆除。 1.2 工作负责人认真检查检修设备及场地无遗留物
			2. 未按要求办理终结手续	2.1 工作负责人确认所有工作班成员已全部撤离现场。 2.2 工作负责人与工作许可人共同验收检修设备已恢复至开工前状态

续附表

序号	工作项目	伤害类型	危险源	控制措施
八	临时用电、厂家现场服务人员、实习人员现场作业	直接:触电伤害、设备损坏	失去监护	(1) 开工前必须要同本单位其他人员一起参加工前交底会议。 (2) 每项工作必须征得工作负责人同意,必须在监护人(负责人、或由其指派的小组负责人)的监护之下进行作业。 (3) 临时用工和实习人员不得独立作业、不得从事技术工作
九	施工电源	直接:触电伤害、设备损坏	1. 电源取用不当	1.1 施工电源的引接必须在站内指定地点且征得运行人员同意后进行。 1.2 交流电源必须设漏电、短路保护装置,其跳闸电流应与上级空气开关或熔断器相配,严禁越过保护装置接电源,严禁插线取电、三相四线检修电源必须使用专用零线
			2. 检修电源板配置不合理	电源容量要满足工作需要
			3. 拆接电源方法不当	3.1 正确接、拆交流电源,核对设备与电源电压是否一致。 3.2 拆接电源时应断开电源开关,并悬挂"有人作业,严禁合闸"警示标志牌,设专人负责监护
			4. 电源线破损、规格不符	4.1 电源线线径、长度、绝缘满足要求。 4.2 电源线路径合理,有防止辗、砸、压措施
			5. 搭接不规范	电源线接头处必须用绝缘胶布包好
十	消防安全管理	直接:火灾	1. 灭火设施配置、使用不当	1.1 现场应有相应合格的灭火器材,位置适当,数量充足。 1.2 工作人员应熟练使用灭火装置
			2. 焊接操作不当(电焊、火焊)	2.1 乙炔瓶必须直立放置,与明火保持10 m以上距离,并采取防曝晒措施。 2.2 电焊人员应正确佩戴防护用品(面罩、防护镜、手套等)。 2.3 氧气瓶和乙炔瓶摆放保持5 m以上距离。 2.4 严格执行动火工作票制度
			3. 滤油设备起火	3.1 所有滤油设备必须可靠接地。 3.2 及时清理使用用过的滤油纸及油布,严禁乱扔 3.3 滤油现场严禁吸烟,严禁使用电焊、火焊。 3.4 现场作业人员应熟悉常用消防器材的使用方法

续附表

序号	工作项目	伤害类型	危险源	控制措施
十一	起重工作	直接：触电伤害、高空坠落、物体打击、设备损坏	1. 起重机具选用不当	1.1 根据被起吊物及环境情况选用合适的起重机具。 1.2 起重机具检验证审查合格
			2. 司吊人员无证操作	司吊人员要具备相应资质
			3. 吊绳选用不当	必须选用合格、匹配的吊绳
			4. 吊钩无防脱钩装置	吊钩应有防止脱钩的保险装置，并保持装置完好
			5. 操作方法不当	5.1 起重机作业前，施吊人和指挥人应相互沟通，并明确旗语、手势或口哨。 5.2 指挥人员发出的指挥信号必须清晰、准确，辅助作业人员必须听从指挥。 5.3 起吊重物时，吊臂及被吊物下严禁站人；必要时，在起吊范围内设置安全围栏。 5.4 起重机在进入起吊作业区时按确定的路径进入并由专人引导。 5.5 在撑腿前要检查着力点，不能在电缆沟或其他不可着力点支撑吊车腿，并保证施吊过程中重心稳定。 5.6 在带电设备周围起吊时与相邻带电设备保持足够安全距离（10 kV，3 m；35 kV，4 m；110 kV，5 m；220 kV，6 m），并设专人监护。 5.7 起重机在施吊过程中应可靠接地。 5.8 吊起的重物不得在空中长时间停留。在空中短时间停留时，操作人员和指挥人员均不得离开工作岗位。 5.9 起吊前应检查起重设备及其安全装置；重物吊离地面约 10 cm 时应暂停起吊并进行全面检查，确认良好后方可正式起吊。 5.10 起吊物应绑牢，并有防止倾倒措施。吊钩悬挂点应与吊物的重心在同一垂直线上，吊钩钢丝绳应保持垂直，严禁偏拉斜吊。 5.11 起重机严禁同时操作三个动作，在接近满负荷的情况下，不得同时操作两个动作。重物吊离地面 10 cm 时，应暂停起吊并进行全面检查，确认良好后方可正式起吊。 5.12 钢丝绳防止打结、扭曲，穿过滑轮的钢丝绳不能有接头，钢丝绳不得与物体的棱角直接接触，应在棱角处垫以半圆管、木板或其他柔软物。 5.13 吊绳的夹角一般不大于 60°，最大不超过 90°

续附表

序号	工作项目	伤害类型	危险源	控制措施
十二	电动机具使用	直接；触电伤害、机械伤害	1. 操作不当	1.1 必须使用合格的电动机具。 1.2 使用者必须具备相应的操作技能，并熟知操作手册。 1.3 现场使用钻床、切割机、砂轮机、手提电钻等严禁戴手套。 1.4 现场使用切割机、砂轮机时必须佩戴护目镜
			2. 设备漏电	2.1 用电设备电源线绝缘良好；外壳良好、可靠接地。 2.2 所取用的电源必须具有漏电、短路保护装置

附表2　推荐的查勘原始记录表格式

工作内容		计划工作时间	
		查勘时间	
内容	现场情况记录确认		执行人
一、收集整理设备运行情况及缺陷情况和家族缺陷通报			
1. 目视检查本次任务设备的状况			
2. 查询设备厂家说明书、场地安装图纸等相关技术资料			
3. 查阅设备运行记录，查看近期有无事故和异常			
4. 查阅设备修试记录和历次试验报告，确认设备状况和遗留缺陷			
5. 查询目前该型号产品有无家族缺陷通报情况			
6. 查询技改要求或反措文件是否有对应项目			
7. 初步分析要实施的具体工作			
8. 收集计划工作时间段的气象资料			
二、查勘本次工作任务需要的现场基本条件			
1. 核对现场电气主接线布置			
2. 确认需要停电的设备、间隔和范围			
3. 确认保留的带电部位			
三、查勘施工位置是否满足要求			
1. 确认需要实施工作的具体部位、位置			
2. 确认人员工作位置与邻近带电设备间的安全距离满足要求			
3. 预估现场所需的主要施工机具和设备			
4. 预估器材摆放位置，确认与工作区域边缘间的安全距离满足要求			

续附表

工作内容		计划工作时间	
		查勘时间	
	内容	现场情况记录确认	执行人
一、收集整理设备运行情况及缺陷情况和家族缺陷通报			
5. 预估起重及登高器材工作位置,确认与邻近带电设备间的安全距离满足要求			
6. 预估动火工作位置,确认与邻近带电设备和易燃品的安全距离满足要求			
7. 预估其他可能威胁到本次施工安全的危险点			
8. 预估现场所需的安全措施			
9. 使用图像记录现场的实际情况			
四、查勘工具、设备进入工作区域的通道是否畅通			
1. 人员、设备、器材到达变电站道路通畅,承重、限高满足要求			
2. 设备、器材进出现场存放位置道路通畅,承重、限高满足要求			
3. 人员、设备、器材进出工作地点道路通畅,承重、限高满足要求			
五、形成查勘记录			
1. 整理本次查勘获得的信息和图像			
2. 编制查勘原始记录,工作负责人和参与查勘的其他人员需要签字			
六、其他情况备注			
查勘人员签字			工作负责人签字

Module 3　Standardized Operation

　　Generally speaking, the on-site standard process of substation maintenance includes three phases, namely, the preparation for on-site operation, the maintenance implementation, and the maintenance completion and summary. Each phase can be divided into several steps.

1. Preparation for On-site Substation Maintenance

1) Getting Work Tasks

(1) Ways to issue a work plan.

Usually, the substation maintenance schedule for a grid system of the State Grid Corporation of China can be roughly divided into three categories, i.e. scheduled maintenance, general defect elimination, and critical fault handling.

Scheduled maintenance (hereinafter referred to as the maintenance schedule) is developed level by level. The competent department of operation and maintenance of equipment (hereinafter referred to as the Equipment Department) is mainly responsible for formulating the substation maintenance schedule. The development of the maintenance schedule usually starts in September each year and the maintenance schedule is issued in December of that year. The content of the maintenance schedule is sourced from maintenance, O & M, materials, planning and users, and many other aspects related to the operation of systems and equipment. In the prefecture-level power supply enterprises under the State Grid Corporation of China, the departments are responsible for planning, coordinating, and assigning the interruption maintenance tasks of the equipment under the management of the corresponding enterprises and customer equipment in the power supply business area.

The Operation and Maintenance of Equipment Department (hereinafter referred to as the Equipment Department) is the department that is responsible for the centralized management of maintenance of the company's power grid transmission and transformation equipment and urban distribution network. It is responsible for organizing, coordinating, and formulating annual maintenance schedules, approving maintenance projects, checking the implementation of maintenance schedules, assessing the safety, progress, and quality of maintenance, and organizing acceptance checks of major maintenance projects.

The company's Dispatching and Control Center (hereinafter referred to as the D & C Center) is responsible for organizing people to review the interruption maintenance schedules submitted by operating organizations. It is responsible for formulating the overall scheme of interruption maintenance schedule of power transmission and transformation equipment in the grid, determining the operation mode of the grid during the maintenance, issuing and implementing the application for dispatching and maintenance, supervising the implementation of the maintenance schedules of power transmission and transformation equipment, collecting statistics on the completion of maintenance of each organization, cooperating with the Equipment Department to make maintenance project arrangements and assess maintenance of power transmission and transformation equipment.

The Marketing (Electricity for Rural Areas) Department of the company is responsible for assisting and urging customers within the scope of responsibility to develop maintenance schedules, reviewing the maintenance work plans to cut off the power direct supplied to the customers, and incorporating them into the annual and monthly interruption maintenance schedules. The Marketing (Electricity for Rural Areas) Department is the department that is responsible for the centralized management of maintenance of the company's power distribution network and equipment in rural areas. It is responsible for organizing, coordinating, and formulating annual maintenance schedules, approving maintenance projects, checking the implementation of maintenance schedules, assessing the safety, progress, and quality of maintenance, and organizing acceptance checks of major maintenance projects.

(2) Process for issuing a maintenance schedule.

The Equipment Department submits the equipment maintenance schedule to the Equipment Department of the provincial company, and the Equipment Department of the provincial company sort and organize the maintenance schedules from all over the province, and submit them to a regional power grid company such as the southwest branch of the State Grid Corporation of China together with the maintenance requirements for the voltage classes over 220 kV. And in a similar fashion, these maintenance schedules accrue each time they are submitted to a higher level until they reach the State Grid Corporation of China. The State Grid Corporation of China coordinates the maintenance schedules of regional power grid companies, and adds and designates some time nodes when cross-regional or key plans are needed. In this way, pre-arrangements of the maintenance schedules are developed. These pre-arrangements are issued to the regional power grid companies, and also after the time nodes of the same work are specified, the pre-arrangements are issued to the provincial companies, the Equipment Departments of the municipal companies, and then to the front-line maintenance organizations. The operation and inspection management personnel of the front-line maintenance organizations shall prepare for the procurement and dispatching of necessary materials according to the specified time nodes, break the remaining tasks without specified time into quarterly schedules and monthly schedules, and issue them to the teams. The team leader will finally divide the monthly schedule into the weekly schedules, and assign the person in charge of work to implement the work. The specific process is as follows:

① The Provincial Dispatching Department is responsible for managing the issuing process of the power interruption schedule of equipment.

Annual schedule: The competent business departments (such as Operation and Maintenance Department, and Construction Department), maintenance management organization, and construction organization are responsible for preparing the power interruption requirements for the next year of the equipment under the management of the Provincial Dispatching Department. After these power interruption requirements are reviewed by the Equipment Department and approved by the competent leaders, they are submitted by the competent dispatching and control department to the Provincial Dispatching Department, and issued after being balanced by the Provincial Dispatching Department.

Quarterly schedule: The Equipment Department, Construction Department, maintenance management organization, and construction organization are responsible for preparing the power interruption requirements for the next quarter of the equipment under the management of the Provincial Dispatching Department according to the annual interruption maintenance schedule. After these power interruption requirements are reviewed by the Equipment Department and approved by the competent leaders, they are submitted by the D&C Center to the Provincial Dispatching Department, and issued after being balanced by the Provincial Dispatching Department.

Monthly schedule: The Equipment Department, Construction Department, maintenance management organization, and construction organization are responsible for preparing the power interruption requirements for the next month of the equipment under the management of the Provincial Dispatching Department according to the annual interruption maintenance schedule. After these power interruption requirements are reviewed by the Equipment Department and approved by the competent leaders, they are submitted by the D&C Center to the Provincial Dispatching Department, and issued after being balanced by the Provincial Dispatching Department.

Weekly schedule: The Maintenance Construction Organization is responsible for submitting the next week's maintenance schedule to the D&C center one week before commencement according to the quarterly and monthly maintenance schedules of the Provincial Dispatching Department, and then the D&C Center shall submit it to the Provincial Dispatching Department.

② The D&C Center of the prefecture-level power supply company is responsible for managing the preparation process of the power interruption schedule of equipment.

Annual schedule: The competent business departments (such as Operation and Maintenance Department, Construction Department, and Marketing Department), maintenance management organization, and construction organization are responsible for submitting the next year's power interruption requirements of their own to the D & C center for summary. The Marketing Department is responsible for collecting the next year's power interruption requirements the big users have for equipment of 35 kV and above, summarizing them, and submitting them to the D&C Center. The D&C Center shall, together with relevant departments, compile the *Annual Maintenance Schedule for Grid Equipment* and submit it to the power supply company for approval.

Monthly schedule: Each of the maintenance and construction organizations shall provide the next month's power interruption requirements to the equipment operation organization. After receiving their power interruption requirements, the equipment operation organization shall not delete any of the power interruption requirements, and different opinions if any shall be put forward at the monthly maintenance balancing meeting of the company. After the initial internal balancing, the equipment operation organization shall submit the power interruption requirement plan for the next month to the D & C Center, Operation and Maintenance Department, Construction Department, Marketing Department, and competent business departments. The competent Marketing Department is responsible for collecting the next month's power interruption requirements of the special line users, summarizing them, and submitting them to the D&C Center. The D & C Center shall comprehensively balance the requirements together with relevant departments and have a monthly maintenance balancing meeting for the company's competent leaders.

After the monthly maintenance balancing meeting of the prefecture-level company, each county-level branch shall submit the official version of the monthly interruption maintenance schedule (signed and sealed by the competent leaders) of their own to the D&C Center of the prefecture-level company. The D&C Center of the prefecture-level company shall summarize and compile the *Monthly Interruption Maintenance Schedule for Grid Equipment* and submit it to the prefecture-level company for approval and issue.

The power interruption requirement plan to be submitted shall be detailed and clear, and it's necessary to specify the scope of equipment to be power interrupted, main work content, power interruption time, and other special requirements.

(3) Mode of issuing a maintenance schedule.

Maintenance schedules are divided into scheduled maintenance and unscheduled maintenance to facilitate management.

Scheduled maintenance refers to equipment maintenance, testing, upgrading, capital construction, and other work requiring power interruption conducted at the time approved by the Dispatching Department, which are incorporated in the annual, monthly, and weekly power interruption schedules,

Unscheduled maintenance refers to maintenance other than scheduled maintenance and is divided into general defect elimination and critical fault handling.

For general defect elimination, when any general defects are unexpectedly found in the routine inspection or operation, the Operation Department or Maintenance Department usually will submit the requirement to eliminate such defects and address them in the next interruption maintenance schedule. Generally, there will be some time to prepare for this type of maintenance task.

Critical fault handling is to deal with any emergent faults arising from equipment operation, which will otherwise seriously threaten the normal operation of the system if not eliminated timely. Generally speaking, such a task shall be carried out by the maintainer on site immediately after receiving the maintenance requirement report from the Operation and Maintenance Department.

In theory, each maintainer can learn about the maintenance schedule of the year from the operation and maintenance management personnel of the organization and learn about the planned maintenance task of the month from the team leader, which enables the maintainer to prepare in advance. In contrast to critical fault handling tasks for which the maintainers have no time to prepare and are unlikely to practice pertinently, scheduled maintenance and general defect elimination are compared to open-book examinations for which maintainers have enough time to prepare and practice pertinently. Such open-book examinations are a process in which maintainers gain experience and progress from their own maintenance practices, while fault handling is a test of the maintainers' maintenance level.

2) Performing Field Investigation

(1) Basis of field investigation.

For substation maintenance (construction), as required by the *Electric Power Safety Working Regulations (Power Transformation)*, the maintenance (construction) organization shall organize field investigation according to the work task and complete the field investigation records if the work ticket issuer or the person in charge of work considers it necessary to make field investigation. The field investigation shall be organized by the work ticket issuer or the person in charge of work.

Although the Regulations suggest that the field investigation can be performed whenever "necessary", as the unattended substations keep scaling up, an investigation is recommended when conventional Class A, B, C, and D maintenance is carried out or when there are any live parts around the work site, to keep abreast of the latest actual situations and thus better prepare for maintenance.

(2) Ways of investigation.

People shall carry out the investigation on site. It is recommend that the investigation shall be performed 7 to 15 days before the scheduled commencement, better not too late or too early. A late investigation will not give maintainers enough time to prepare. Besides, a power interruption application shall be filed at least one week earlier for work requiring power interruption to leave the Dispatching Department enough time to transfer loads and give power interruption notice. For example, if the power interruption application is submitted 7 days in advance, the approval takes a minimum of 2 days, a work ticket requires 1 day earlier, and the maintainers are really pressed to complete scheme preparation, audit, material preparation, and learning in the last four days. It's not advisable to file the application too early because there may be any investigation information deviations due to its time sensitivity.

(3) Investigation content.

The *Electric Power Safety Working Regulations (Power Transformation)* does not provide for what the investigation is about. Here we quote Article 5.2.2 of the Regulations on lines: For on-site investigation, it's required to check the scope of power interruption for the on-site construction (maintenance), the retained live parts, and the conditions, environment, and other hazards of the work site. In work, there are many details that need to be investigated, which can be referred to Schedule 2.

3) Preparation of Construction Scheme

According to the results of the on-site investigation, organizational measures, technical measures, and safety measures shall be formulated for the work items with greater risk, complexity, and difficulty, and shall be implemented after approval by the leader (chief engineer) in charge of the power generation of the organization.

Content of construction scheme: The construction scheme is the core of the construction preparation, the mental rehearsal and preparation basis of maintenance. It is also a written basis for the division of labor, maintenance content and technical standards for the work, and is the logical representation of the construction process and schedule. Generally, the construction scheme is prepared as a document by the person in charge of work. Although it's not written in a uniform format, its content can be roughly classified as organizational measures, technical measures, and safety measures. The maintenance scheme is thus known as the "three kinds of measures". Organizational measures are mainly for allocating resources related to "people", technical measures are all about allocating resources related to "things", and safety measures are aimed at adjusting the above two to ensure safety.

The three kinds of measures are independent but interrelated. They complement each other logically, and in terms of viewpoint, they are against each other, which will always be the case throughout the preparation of them. Together, they make the construction scheme complete, correct, and objective.

The organizational measures mainly need to solve the following four problems: What are the positions for the task? What are the technical and safety responsibilities of each position? How many people and what skills are required for each position? Who are the best candidates for these positions?

Technical measures mainly need to solve the following problems: What's the order of the specific content of the maintenance task? What are the technical and quality control standards? What materials are needed to support technical and quality control? What is the schedule for on-site implementation? How to arrange the construction site? Is the actual cost estimation consistent with the budget?

4) Approval of Standardized Operation Documents

For all kinds of standardized operation documents used in on-site operations, corresponding approval procedures must be handled and the tracking responsibility system must be implemented. When approving the standardized operation documents, the whole operation process must be comprehensively reviewed, focusing on reviewing hazard analysis, control measures, working procedures, etc. After approval, the approving officer must sign and comment (including suggestions on revision).

(1) Approval principle: Standard cards shall be approved by people at different levels according to the organizational relationship of operation management.

① The standard cards for major repairs of power transmission lines of 110 kV and below and auxiliary equipment, and conventional Class C and D maintenance and routine inspections of power transmission and transformation equipment of 220 kV and below shall be reviewed by the relevant technicians of the specialty workshop (room) and approved by the supervisor (deputy supervisor).

② The construction organization schemes or standard cards for major repairs of 35 kV and 110 kV main transformers, 220 kV auxiliary equipment and 220 kV power transmission lines shall be reviewed by the Specialist Engineers of the specialty workshop (room) and the Operation and Maintenance Department and submitted to the (deputy) director of the Equipment Department for approval.

③ The construction organization schemes or standard cards for technical upgrading of 35 kV and 110 kV equipment shall be reviewed by the Specialist Engineers of the specialty workshop (room) and the Operation and Maintenance Department and submitted to the (deputy) director of the Equipment Department for approval.

④ The construction organization schemes or standard cards for major repairs of 220 kV main transformers, technical upgrading of 220 kV main equipment, or other special and large construction tasks determined by the Equipment Department shall be reviewed by the specialty workshop (room) and the Equipment Department, jointly heard by the Safety Supervision Department, and submitted to the deputy general manager of power generation (chief engineer) of the power supply company for approval.

(2) Learning of standardized operation documents.

The team leader shall organize a pre-shift meeting before the operations to learn all the approved operation documents one by one.

2. Operation on the Substation Maintenance Site

After making all the preparations for the maintenance, the person in charge of work shall, by the approved maintenance time, bring the standardized operation flow chart, work ticket and safety learning records prepared in advance and lead the work team members to the work site. Before departure, the person in charge of work shall reconfirm the weather, the availability of materials, and the attendance of members.

In general, when the dispatcher is approving the maintenance time, the maintenance is usually will scheduled for the morning to ensure sufficient time for maintenance and avoid the disadvantages and safety hazards brought by night construction. The power interruption and switching operations of the equipment under maintenance by the O & M personnel and the safety measures are usually scheduled for after midnight of the previous day. At present, due to the integration of operation and maintenance, the personnel and time of the power interruption may vary from organization to organization. If the power interruption is delayed and the O & M personnel fail to finish providing the safety measures after the maintainers arrive, the maintainers will be required to wait and not go anywhere. In this case, the maintainers shall not stay in the substation in theory, which isn't strictly followed at work. However, the maintainers shall stay together and must not enter the substation equipment site or the central control room. After the O & M personnel take all the safety measures, the permitter shall contact the person in charge of work to apply for the work permit.

1) On-site Work Permit Procedures

When it comes to handling the on-site work permit procedures, the work permitter shall, together with the person in charge of work, confirm that the safety measures listed on the work ticket have been taken properly on site, and the work permitter shall explain clearly the scope of power interruption to the person in charge of work and prove that the equipment is indeed power-interrupted in person. Both parties will sign the work ticket in person if they have no doubts, and the person in charge of work and the permitter will each hold a work ticket and return to their respective posts. The person in charge of work shall bring the work ticket and lead the work team members to make preparations on site before commencement. The work permitter shall keep the work ticket and write down the permitted start time of work on the operation record book.

In particular, it shall be noted that the work team members can enter the site only if they have handled the on-site work permit procedures and are led by the person in charge of work.

2) Convening a Pre-shift Meeting

As the on-site work permit procedures are only handled by the person in charge of work, without the participation of the work team members, the person in charge of work shall convene a pre-shift meeting to learn about safety from the approved operation documents. The person in charge of work needs to inform every member of the work team of this meeting and make sure that they understand. This is also known as the disclosure procedure of on-site safety technical measures. The form of the pre-shift meeting is relatively unified. In most cases, the work team members would line up empty-handed on the site, and the person in charge of work first would check the clothes and mental state of the members and ask them if they are ok or if they have any requirements. Then the person in charge of work would make a presentation, which shall at least clearly explain four things: the operation flow of the task; the division of labor; the safety measures, hazards, and control measures of the work site; the scope of power interruption of the work site; and the retained live parts. The work team members can ask questions during the presentation. The person in charge of work shall ask random members questions after the presentation to ensure that all work team members get the information.

After confirming the tasks assigned by the person in charge of work and the safety measures of the construction project, the work team members shall sign the work ticket held by the person in charge of work for confirmation. Only the work team members who have signed can go on to work.

When a secondary work ticket is adopted, the person in charge of the divided work shall sign the general work ticket, and the other work team members shall sign the secondary work ticket.

3) Implementation of On-site Operations

The implementation stage of on-site operations is actually the stage of carrying out the maintenance scheme. Maintainers shall be divided according to the organizational measures, perform the process and standards of the technical measures, and strictly follow the requirements of safety measures to avoid risks. The main task of the person in charge of work is to direct, supervise and confirm the process and quality while the work team members are responsible for specific operations.

How can a technical solution laden with details be orderly implemented on site? A standardized operation flow chart will do the work. A standardized operation flow chart can be regarded as a version of the maintenance scheme on implementation and shall be carried by the person in charge of work like a work ticket. A standardized operation flow chart is for reminding and reference. Because there are no process or document requirements, the maintainers must rely on their memory to construct during maintenance. Generally speaking, any maintenance task included in the work plan has a minimum of one week for preparation, which is similar to an open-book examination for the maintainers. The maintainers' expertise is tested on site only in addressing any emergencies or defects when a detailed maintenance scheme cannot be prepared in advance due to time limits.

Therefore, the person in charge of work and the work team members shall make full use of the standardized operation flow chart as a reminder and reference in the operation, and direct their energy and attention from extracting the on-site process and standards to the implementation and control of the operation. In short, they shall put down what's correct and definite on paper in advance and follow what the card says when on site. What they shall pay attention to on site is whether they have done it correctly and safely.

The composition of the standardized operation flow chart is in line with the working procedures, and the operating steps consist of the operating items in the technical standards. Each operating step shall be executed cyclically in accordance with the same cycle, and the cycle is divided into four links:

Issue: The person in charge of work shall give instructions and read the technical standards and hazards one by one in the sequence of items stated in the standardized operation flow chart.

Implementation: After repeating the instructions, the work team members shall perform operations in strict accordance with the methods and processes of the technical standards.

Verification: After the work team members complete the operations, the person in charge of work shall check by referring to the verification plan in the technical standards.

Confirmation: After the person in charge of work checks the parameters and keeps records, the operator shall sign in the maintainer column of this step on the operation flow chart. The person in charge of work shall check and sign in the confirmation column of this step on the operation flow chart. This step comes to an end, and the cyclical issue link of the next step begins. If any original records such as the test report printouts arise from this step, the operator and the reviewer need to sign and indicate the time on the original copies. If the record is an electronic file, the person in charge of work shall indicate the file name and storage address in the operation flow chart. It is recommended that the person in charge of work use an independent removable storage device for on-site backup.

In the operation, in case of any inconsistencies with the situation on the site, relevant drawings, and relevant provisions, the person in charge of work shall timely modify the process card according to the site situation, and if necessary, inform the original approver and seek his/her approval before continuing the operation. After the operation, the original approver shall make up for the signing procedures.

In the operation, if the equipment is found to have defects or anomalies that have not been discovered beforehand, the operator shall immediately report to the person in charge of work (the group leader), analyze them in detail, and propose solutions. After addressing the defects and anomalies, the operator shall truthfully complete the corresponding columns of the on-site operation flow chart one by one. The person in charge of work (the group leader) must sign in the "confirmed by" column upon confirmation and acceptance, and it is strictly forbidden to simply mark "√".

Any defects and anomalies that cannot be handled shall be truthfully taken down one by one in the on-site operation flow chart and signed in the "discovered by" column for confirmation. The person in charge of work (the group leader) shall sign for confirmation in the "confirmed by" column.

When the equipment cannot be put into operation normally due to defects and anomalies, the chief in charge of work shall inform the original approver and seek his/her approval before continuing the operation. After the operation, the original approver shall make up for the signing procedures.

In the operation, if the operator has to change the operation flow or operation items due to any contingencies, the operator shall immediately report to the person in charge of work (the group leader). The person in charge of work (the group leader) shall advise on changes and adjustments in time, and complete the "operation flow" and "changes to operation items" columns of the on-site operation flow chart one by one, and the discoverer and the person in charge of work (the group leader) shall sign for confirmation at the "applied by" one by one. If necessary, inform the original approver and seek his/her approval, and adjust the operation content before proceeding with the operation. After the operation, the original approver shall make up for the signing procedures.

Throughout the process, the on-site operators shall do exactly as the documents prepared in advance tell them to, and in principle, they need not and cannot make any changes without authorization. Fundamentally, to avoid changes is not to reject thinking and innovations, but to say no to work that is not well planned or designed in advance, lest mistakes be caused because of uncertainty. Therefore, the objects that "can not be changed" refers to the approved plans, verification, audit programs, operation flow, processes, and standards. They are only limited to the implementation links of on-site operations. Therefore, if ignoring or disregarding the meticulous, rigorous, complete, and operable planning and preparations in preparing the maintenance scheme, the maintainers will inevitably face the dilemma that there is no basis or the basis is too general to be implemented in the operation, which is the source of confusion for maintenance per se. In this case, the people's safety and the maintenance quality are simply out of the question.

3. Completion of On-site Substation Maintenance

1) Keeping the On-site Maintenance Records

When it comes to maintenance and testing, according to the vertical comparison analysis principle for condition-based maintenance that is spread over three cycles in three years, the maintainers shall take down the original maintenance data in detail and properly save it as the basis for future maintenance and important traceability information. Besides, it is necessary to go to the on-duty O & M personnel in the substation control room to keep the maintenance and testing records. What shall be taken down include the content of the maintenance task, what's addressed, and the remaining problems. At the end of the records, it's required to clearly describe the maintenance result as to whether the maintenance task is completed and whether the equipment concerned can be put into operation.

2) Performing Three-level Acceptance

After the maintainers complete all operations in accordance with the process and content of the standardized operation flow chart and verify that the operations are correct, the maintained equipment shall be put back to its original position and set to the initial state before the maintenance, the work site shall be cleaned, and the safety measures established by the work team shall be removed by the team.

Cleaning of work site: Waste shall not be discharged at will, especially not discharged in the substation's drainage ditches to prevent pollution of the water system outside the substation. Instead, it must be recycled and treated in accordance with the provisions. The floor shall be cleaned, and any oil stains shall be washed off. The equipment and materials shall then be transferred to the entrance of the fence, but shall not be loaded onto any vehicle, in order to tackle any problems found in the coming three-level acceptance.

For power transmission and transformation equipment, a two-level acceptance system is adopted for the annual inspections, preexaminations, and maintenance, and a three-level acceptance system is implemented for the major repairs and technical upgrading.

The conditions for applying for acceptance are:

(1) All the maintenance items and content are completed.

(2) The quality of maintenance meets the maintenance standards.

(3) The maintenance records, test reports, and signatures are complete.

Therefore, before the formal application for acceptance, the person in charge of work shall request the team to inspect the maintenance by themselves, focusing on the completion of the above three points.

The first level of acceptance refers to self-inspection by the maintenance organization. The professional technical supervisor of the maintenance organization shall be responsible for acceptance as the manager of the maintenance. The person in charge of work shall file an application to the competent technical department of the maintenance organization (Equipment Department). Usually, it is the technical supervisor of operation and inspection or an officer with specific responsibility who is responsible for on-site acceptance. In addition to the quality of maintenance, it is also necessary to pay attention to the implementation of the maintenance task and the remaining problems of the equipment, so as to support the arrangement of the next power generation task.

The second level of acceptance refers to the acceptance of equipment by the operation organization. The professional technical supervisor of the maintenance organization shall apply to the competent technical department of the operation organization for acceptance. Upon arrival, the technical supervisor of the operation organization shall accept together with the permitter and the on-duty O & M personnel the same day. This level of acceptance is different in that it shifts the focus from maintenance to use. As the user, the operation organization will pay more attention to the remaining faults of the equipment, remote local operation, state indications, signal feedback and transmission of a whole group, and other functions related to the operating state.

The third level of acceptance is the acceptance organized by the power supply company. It is usually carried out by relevant professional technical supervisors of the power supply company simultaneously with the second level of acceptance.

An (on-site) signature responsibility system and a quality traceability system shall be implemented for the acceptance of all projects. If any non-conformances are found during acceptance, a non-conformance notification will be issued and corresponding procedures will be implemented to address them.

3) Handling End-of-work Procedures

After the acceptance, the maintainers can load the tools and instruments they bring with onto the vehicle, and the vehicle shall return to the designated position as soon as possible after loading.

The person in charge of work will convene a post-shift meeting for all the work team members anywhere other than the equipment area, which echoes the pre-shift meeting. The members also need to line up, and the person in charge of work shall preside over the meeting. The post-shift meeting is mainly about verification that the operations are correct, the equipment under maintenance shall be put back to its original position and set to the initial state before the maintenance, the work site shall be cleaned, and the safety measures shall be removed by the work team that has taken them.

Cleaning of work site: Waste shall not be discharged at will, especially not discharged in the substation's drainage ditches to prevent pollution of the water system outside the substation. Instead, it must be recycled and treated in accordance with the provisions. The floor shall be cleaned, and any oil stains shall be washed off. The equipment and materials shall then be transferred to the entrance of the fence, but shall not be loaded onto any vehicle, in order to tackle any problems found in the coming three-level acceptance.

For power transmission and transformation equipment, a two-level acceptance system is adopted for the annual inspections, preexaminations, and maintenance, and a three-level acceptance system is implemented for the major repairs and technical upgrading.

The conditions for applying for acceptance are:

The work shall be deemed terminated when the work permitter stamp "Executed" on the work ticket held by the person in charge of work upon signing by both parties. After the work, the equipment and the site shall be regarded as live equipment and site, and no one shall enter the work site again, let alone work without authorization. The work ticket held by the permitter shall be given to an operator to continue handling the procedures for terminating the work ticket.

The work ticket held by the person in charge of work shall be brought back to the maintenance organization and kept for one year.

Now the maintainers can leave the substation. To ensure normal power transmission by the maintained equipment, the substation maintenance and relay protection disciplines shall arrange for people (usually person in charge) and tools and instruments to be in the substation till the equipment is put into operation. In this process, the people staying there shall behave as if the work permit procedures are not handled.

4. Summary and Evaluation after On-site Maintenance

What shall be summarized and evaluated include safety, quality, items, man-hours, materials and spare parts, technical supervision, cost, maintenance and etc.

The construction organization shall submit the completion report, technical summary, test report, and acceptance report in time.

The maintenance archives shall be sorted out in accordance with the management regulations of maintenance and technical upgrading projects and submitted to the company's archives room for unified archiving. In addition, complete maintenance archives shall be submitted to the operation organization. The maintenance organization is required to keep the archives for one maintenance cycle.

In the event of any accidents due to poor maintenance quality and lenient acceptance, the maintainers and the acceptors concerned shall be held responsible.

Analyze and take stock of the abnormal conditions of equipment, the experience, and lessons learned from maintenance and management, so as to continuously improve the maintenance quality and management level, and improve the reliability of equipment.

After the annual maintenance task of power transmission and transformation equipment, each organization shall statistically analyze and take stock of the completion of the annual maintenance schedule every November. Such statistics and summaries shall be reviewed by the organization's competent leaders and then submitted to the company's Equipment Department and D & C Center. Besides, the maintenance officers of companies at the prefecture level and above can check the completion, evaluation, and closed loop of the annual maintenance schedule on PMS.

Specifically, the person in charge of work shall give the work ticket he/she held to the technical officer within 3 days for keeping. The person in charge of work shall fill in the maintenance records, including the work content, start and end time, place, personnel, identified problems, solutions, and results. The maintenance records are proof of the on-site workload of the work team and will also be used as paper records that the team can refer to for the equipment maintenance records. The person in charge of work shall take the lead in preparing the maintenance report, organize the test results and test data collected from the process data of this work according to the standard operation procedures that are confirmed item by item, and enter and retain them in the specified format. The original copies of test reports and test data generated in the work shall be attached to the paper report and submitted to the technical officer for filing. These materials are also one of the measures of building a star-rated team.

Schedule 1   Identification of Hazards in Substation Maintenance of Typical Equipment and Preventive and Control Measures—General

S/N	Working item	Type of damage	Source of hazard	Control measure
1	Site survey	Direct: electric shock Indirect: equipment damage	1. The survey is badly organized.	1.1 The site survey for large-scale interruption maintenance shall be organized by the person in charge of the production technology department of the maintenance organization, and the person in charge of work, the person in charge of each discipline group, and the person in charge of site safety must participate. The site survey for other interruption maintenance shall be organized by the person in charge of work, and the persons in charge of the relevant discipline groups shall participate. 1.2 The surveyors must have appropriate safety knowledge and skills.

Continued

S/N	Working item	Type of damage	Source of hazard	Control measure
1	Site survey	Direct: electric shock Indirect: equipment damage	2. Touch live equipment by mistake.	2.1 Specify the scope of the survey and maintain a sufficient safe distance from live equipment. 2.2 It's strictly prohibited to move, climb over or open the fence.
			3. The survey is not conducted properly.	3.1 On-site surveyors shall learn about the task and its specific working items in advance to avoid increasing the workload temporarily. 3.2 Carefully consult the drawings and technical records. 3.3 The person in charge of work and the team leader must check any abnormal conditions of the equipment under maintenance. 3.4 The person in charge of work and the team leader must check the defect records and operation records of the previous year and the current year. 3.5 The working conditions and work scope of the work site must be properly surveyed, the power interruption scope must be verified, and the content and layout requirements of the site safety measures to ensure work safety must be clearly specified. 3.6 Surveyors must survey the maintenance site in person, and keep survey records. The survey activities should be recorded (sound or pictures recorded). 3.7 A site survey is strictly prohibited in the event of strong thunder, storm, rainstorm, and other inclement weather. The survey shall be stopped immediately in case of bad weather. 3.8 It's strictly prohibited for a person to survey alone, nor should a person stay alone in the HV room and HV equipment area. 3.9 The SF_6 equipment room and capacitor room must be ventilated according to the provisions before entering. Oxygen content and harmful gases must be detected before entering an environment that is closed for a long time, such as a cable trench or a cable shaft. 3.10 Do not attempt to operate all electrical equipment in use during the survey, and do not pull the terminal board wiring in operation. Avoid touching exposed LV cables, terminal blocks, and connecting plates by mistake, lest the secondary air switch trip due to accidentally touching.
			4. The site is poorly lit.	The site under survey shall be well lit and if necessary, a movable lighting facility shall be used.

Continued

S/N	Working item	Type of damage	Source of hazard	Control measure
2	Preparation of machinery, safety tools and instruments, and materials	Direct: mechanical injury, fire Indirect: equipment damage	The machinery is not qualified, not well prepared, and not handled in accordance with the provisions.	1.1 When handling large or heavy equipment, do not directly carry it on the shoulder, use ropes and carrying poles. If an object needs to be carried by many people, a person shall be specially designated to give commands, signals shall be unified, and people shall carry it at the same pace. 1.2 When transporting equipment on a rainy or snowy day, take antiskid measures. When carrying the equipment on a steep slope, set up anti-slip facilities on the road while reducing the weight that each person carries. 1.3 The lifting tools used for transportation shall be sturdy and reliable and shall be checked by the person in charge of work each time they are used. The mildewed rope shall not be used. 1.4 When loading and unloading a rolling object with a springboard or log, control it with a rope, fix the remaining items, and no one is allowed to stand in front of where the object is going to roll. 1.5 When using a crane for loading and unloading, the safety regulations of lifting machinery must be observed. 1.6 The machinery shall be in good operating condition, safety tools and instruments shall be qualified, and their models and quantity shall meet the work requirements. Check and confirm before packing. Do not use defective machinery, tools, and instruments. 1.7 Strictly implement the machinery management system, and regularly overhaul and maintain the machinery. 1.8 Inflammable and explosive materials, hazardous chemicals, and toxic objects shall be handled and stored in accordance with safety regulations.

Continued

S/N	Working item	Type of damage	Source of hazard	Control measure
3	Transportation	Direct: Personal injury or equipment damage	1. Someone drives without a license.	Strictly examine the qualifications of drivers.
			2. Drive a faulty vehicle	2.1 It is strictly forbidden to drive a faulty vehicle. 2.2 In the high-temperature season, focus on checking whether the tire temperature and pressure are normal.
			3. The vehicle is overloaded, ultrahigh, super-wide, or travels above the speed limit.	3.1 An ultrahigh and super-wide vehicle shall be guided into the substation by a specially designated person. 3.2 To transport equipment, wire reel, materials, and other objects that are likely to roll (glide), they must be tied firmly to prevent rolling to and fro. To transport super long and ultrahigh materials, clear marks shall be provided, and the time, route, and speed specified by the traffic management department shall be abided by.
			4. The anti-vibration measures are not suitable.	Take appropriate anti-vibration measures.
			5. Drive under the influence of fatigue, alcohol, and illicit drugs	Drivers are strictly prohibited from driving any vehicle under the influence of fatigue, alcohol, and illicit drugs.
4	Confirmation of safety measure	Indirect: personal injury	1. The work ticket is not acceptable.	1.1 The work ticket issuer and the person in charge of work shall carefully fill in the work ticket. 1.2 The work ticket issuer and the work permitter shall carefully review the work ticket.
			2. The on-site safety measures are inadequate or not correct.	The person in charge of work and the work permitter shall confirm that the safety measures are correct and adequate before handling the permit formalities.

Continued

S/N	Working item	Type of damage	Source of hazard	Control measure
5	Pre-commencement disclosure	Indirect: Personal injury or equipment damage	1. No pre-commencement disclosure is carried out.	The re-commencement disclosure must be carried out before the commencement.
			2. The pre-commencement disclosure is not clear.	The person in charge of work shall organize the work team members to learn the work ticket and have a safety and technology disclosure. All people shall be clear about four things (the task, hazards, operation flow, and safety measures).
			3. The work team members are not clear about the four things.	3.1 Ask questions randomly as to the four things that they should be clear about. 3.2 The work team members shall ask what they are not clear about.
6	Work interruption, and transfer	Hazard: electric shock	1. The safety measures are not properly re-checked.	Resumption of work shall be approved by the person in charge of work, the work ticket shall be retrieved, and the person in charge of work shall re-check if the safety measures meet the requirements of the work ticket.
			2. Enter a compartment by mistake.	2.1 Workers shall not enter the work site if not led by the person in charge of work or a specially assigned supervisor. 2.2 The person in charge of work shall inform the workers about the electrification scope, safety measures, and precautions when transferring to another work site.
7	End of work	Indirect: equipment damage, grid accident, and electric shock	1. The temporary safety measures are not removed.	1.1 The person in charge of work must carefully check if the temporary safety measures are completely removed. 1.2 The person in charge of work must carefully check if there are any residues on the equipment under maintenance and the site.
			2. The end-of-work procedures are not handled as required.	2.1 The person in charge of work shall confirm that all work team members have left the site. 2.2 The person in charge of work and the work permitter shall jointly check if equipment under maintenance is restored to the state before the commencement and accept it.

Continued

S/N	Working item	Type of damage	Source of hazard	Control measure
8	On-site operations for temporary power supply by the on-site service personnel and interns of the manufacturer	Direct: electric shock, equipment damage	There is no supervision any more.	(1) Before commencement, attend the pre-commencement disclosure meeting together with other people from the organization. (2) Each task must be approved by the person in charge of work and must be carried out under the supervision of the supervisor (the person in charge, or a team leader appointed by him/her). (3) Gig workers and interns shall not work alone, nor shall they be responsible for the technical work.
9	Construction power supply	Direct: electric shock, equipment damage	1. The power supply is not used properly.	1.1 The construction power supply must be connected to a designated location in the substation with the consent of the operators. 1.2 An AC power supply must be equipped with leakage and short-circuit protection devices, with the tripping current matching the upstream air switch or fuse. It is strictly prohibited to connect the power supply without using any protection devices, and it is strictly prohibited to insert a cable for power. A special neutral line must be used for a three-phase four-wire maintenance power supply.
			2. The power board provided for maintenance is not suitable.	The power capacity shall meet the needs of work.
			3. The power supply is not connected and disconnected properly.	3.1 Properly connect and disconnect the power supply, and check if the equipment voltage is consistent with the power voltage. 3.2 When connecting and disconnecting the power supply, turn off the power switch, hang a sign board saying "Working! No switch closing!", and specially designate a person for supervising.
			4. The power cord is damaged or inconsistent with the specifications.	4.1 The power cord shall meet the requirements in terms of diameter, length, and insulation. 4.2 The route of the power cord shall be reasonable, and measures shall be taken to prevent it from rolling and being smashed and pressed.
			5. The power cord is not properly joined in an overlapping way.	The connector of the power cord must be wrapped with insulating tape.

Continued

S/N	Working item	Type of damage	Source of hazard	Control measure
10	Fire safety management	Direct: fire	1. Appropriate fire extinguishing facilities are not provided, or fire extinguishing facilities are not used properly.	1.1 Enough appropriate and qualified fire-fighting equipment shall be provided at the right locations of the site. 1.2 Workers shall be proficient in using fire-fighting equipment.
			2. The welding operation is improper (electrowelding, and fire welding).	2.1 Keep the acetylene cylinder upright, keep at least 10 m away from the open flame, and take measures against solarization. 2.2 The welder shall wear protective equipment (masks, goggles, gloves, etc.) properly. 2.3 Keep the oxygen cylinder at least 5m away from the acetylene cylinder. 2.4 Strictly implement the hot work ticket system.
			3. The oil filtering equipment is on fire.	3.1 All the oil filtering equipment must be grounded reliably. 3.2 Clean up the used oil filter paper and oil cloth in time. Do not litter. 3.3 It's strictly prohibited to smoke and use electrowelding or fire welding on the oil filtering site. 3.4 On-site workers shall acquaint themselves with how to use common fire-fighting equipment.
11	Lifting	Direct: electric shock, high-altitude falling, object strike, equipment damage	1. The lifting equipment selected is not appropriate.	1.1 Select the appropriate lifting equipment according to the object to be lifted and the environmental conditions. 1.2 The inspection certificate of lifting equipment shall be checked.
			2. The lifting operator has no certificates.	The lifting operator shall have the required qualifications.
			3. The lifting rope selected is not appropriate.	Select a qualified and matched lifting rope.

Continued

S/N	Working item	Type of damage	Source of hazard	Control measure
11	Lifting	Direct: electric shock, high-altitude falling, object strike, equipment damage	4. The hook is not provided with an anti-slip device.	The hook shall be provided with an anti-slip safety device, and the safety device shall be intact.
			5. The crane is not properly operated.	5.1 Before operating the crane, the lifting operator and the commander shall communicate with each other and clarify the flag signals, gestures, or whistles. 5.2 The commander shall give clear and accurate commanding signals, and the supporting operators must obey the commands given. 5.3 When lifting heavy objects, no one shall stand under the jib or loads. If necessary, safety fences shall be set up within the lifting range. 5.4 The crane shall enter the lifting operation area from the determined path and shall be guided by a specially-assigned person. 5.5 Before propping up the legs, check the acting points. Do not prop up the crane legs in a cable trench or at any other points where force can not be applied. Make sure that the center of gravity is stable during lifting. 5.6 Keep a safe distance from the adjacent live equipment when lifting around it, and specially designate a person for supervising. (10 kV, 3 m; 35 kV, 4 m; 110 kV, 5 m; 220 kV, 6 m). 5.7 The crane shall stay reliably grounded during lifting. 5.8 The lifted weight shall not stay in the air for a long time. When it stays in the air for a short period of time, the operator and the commander shall not leave the operating posts. 5.9 Before lifting, check the lifting equipment and its safety devices. When the weight is lifted about 10 cm above the ground, stop and have a comprehensive inspection. Start lifting formally after confirming that it's OK.

Continued

S/N	Working item	Type of damage	Source of hazard	Control measure
11	Lifting	Direct: electric shock, high-altitude falling, object strike, equipment damage	5. The crane is not properly operated.	5.10 The load shall be tied firmly and measures shall be taken to prevent it from tipping. The hook hanging point and the center of gravity of the load shall be on the same vertical line, and the hook wire rope shall be vertical. It's strictly prohibited to lift any load slantwise. 5.11 It's strictly prohibited to operate the crane to make three movements simultaneously. When it's close to full load, it's not allowed to operate the crane to make two movements simultaneously. When the weight is lifted about 10cm above the ground, stop and have a comprehensive in, spection. Start lifting formally after confirming that it's OK. 5.12 Wire ropes shall be free of knots and twists. Wire ropes threading through the pulleys shall have no joints. Wire ropes shall not be in direct contact with the edges and corners of an object, so the edges and corners shall be padded with semicircle pipes, planks, or other soft materials. 5.13 The included angle of the lifting rope is generally not more than 60°, but not exceeding 90° to the maximum.
12	Use of electric machinery	Direct: electric shock, mechanical injury	1. Misoperation	1.1 Qualified electric machinery must be used. 1.2 The user must have the appropriate operating skills and be familiar with the operation manual. 1.3 It is strictly forbidden to wear gloves when using a drilling machine, cutting machine, grinding machine, electric hand drill, etc. on site. 1.4 Goggles must be put on when using a cutting machine or grinding machine on site.
			2. Electric leakage of equipment	2.1 The power cord of electric equipment shall be well insulated. The housing shall be in good condition and well grounded. 2.2 The power supply used must be provided with leakage and short-circuit protection devices.

Schedule 2   Recommended Format of Original Survey Record Sheet

Work content		Planned working hours	
		Survey time	
Content		Recording and confirmation of site conditions	Performed by
I. Collect and sort out the operation and defects of equipment and family defect circulars.			
1. Visually check the state of the equipment for this task.			
2. Consult the equipment manufacturer's manual, site installation drawings, and other related technical data.			
3. Consult the equipment's operation records to see if there are any recent accidents and anomalies.			
4. Consult the equipment's maintenance and testing records and previous test reports to confirm equipment condition and remaining defects.			
5. Check if this product model has any family defect circulars.			
6. Check if there are any corresponding items in the technical upgrading requirements or anti-accident measure documents.			
7. Preliminary analyze the specific work to be done.			
8. Collect meteorological information for the planned working period.			
II. Survey the basic site conditions required for this work task.			
1. Check the layout of the main electrical wiring on site.			
2. Confirm the equipment, compartments, and scope to which the power supply shall be interrupted.			
3. Confirm the live parts that shall be retained.			
III. Check if the construction location meets the requirements.			
1. Confirm the specific parts and positions that need to be worked on.			
2. Confirm that the safety distance between the working position of the operator and the adjacent live equipment meets the requirements.			

Continued

3. Identify the major construction machinery and equipment that is needed on site.		
4. Identify where the equipment shall be located, and confirm that the safety distance from the edge of the work area meets the requirements.		
5. Identify where the lifting and climbing equipment will be operating, and confirm their safety distance from the adjacent live equipment meets the requirements.		
6. Identify where hot work will be carried out, and confirm that the safe distance from the nearby live equipment and flammables meets the requirements.		
7. Identify other hazards that may endanger the construction.		
8. Identify safety measures required on site.		
9. Take pictures of the situation of the site.		
IV. Check if tools and equipment have unimpeded access to the work area.		
1. The road to transport personnel and equipment to the substation is unimpeded, and the requirements of load-bearing and height limit are met.		
2. The road to transport equipment from and to the storage place on site is unimpeded, and the requirements of load-bearing and height limit are met.		
3. The road to transport personnel and equipment from and to the work site is unimpeded, and the requirements of load-bearing and height limit are met.		
V. Compile survey records.		
1. Sort out the information and images obtained in the survey.		
2. The original survey records shall be compiled, which shall be signed by the person in charge of work and other surveyors.		
VI. Remarks about other conditions.		
Surveyors (signature)		Person in charge of work (signature)

项目三 电气主接线的设计

电气主接线是发电厂、变电站设计的首要部分，也是构成电力系统的重要环节。本书项目一介绍了电气主接线的基本形式与特点，本项目以电气主接线的设计为中心，介绍发电厂、电站电气主接线设计的基本要求、设计程序和步骤，并对主变压器的选择、限制短路电流的方法进行分析。

模块一 设计的基本要求

电气主接线代表了发电厂、变电站高压、大电流的电气部分主体结构，是电力系统的重要环节。主接线的确定与电力系统整体运行的可靠性、灵活性、安全性以及经济性密切相关，同时对电力系统继电保护、自动装置和设备选型、配电装置的布置也有决定性的影响。因此，主接线设计必须综合考虑各方面因素，充分论证比较，最终得到符合实际的最佳方案。

电气主接线设计的基本要求包含可靠性、灵活性和经济性三个方面。

一、可靠性

保证供电的安全可靠是电力生产的首要任务，是电气主接线最基本的要求。停电不仅会使发电厂遭受损失，而且会对社会带来或多或少的经济损失。在经济发达地区，故障导致停电可能会造成设备损坏、产品报废、铁路系统瘫痪、城市生活陷入混乱，甚至出现人身伤亡，造成巨大的经济损失、社会问题和安全问题。因此，主接线形式必须保证供电可靠。

主接线是否可靠不是绝对的，同样形式的主接线对于某一些发电厂、变电站来说是可靠的，但是对于另外一些发电厂和变电站则不一定满足可靠性要求。所以，在分析主接线可靠性时，要充分考虑该发电厂、变电站在系统中的作用、地位，以及用户负荷性质类别、设备制造水平和运行经验等诸多因素。

1. 发电厂、变电站在电力系统中的地位和作用

各发电厂、变电站的电气主接线应当与其接入的电力系统相适应。

大型发电厂或枢纽变电站在电力系统中供电量大、范围广、地位重要，发生事故可能破坏整个系统的稳定运行，造成巨大损失。为此，大型发电厂或枢纽变电站应采用可靠性很高的主接线形式。大型发电厂和枢纽变电站通常距离负荷中心较远，须使用较高电压，容量也很大，因此与系统接入方式宜采用双回路或环网等强联系形式接

入，确保相应电压等级接线方式的可靠性。在设计时，对主接线可靠性需要进行定性分析和定量计算。

中、小型发电厂、变电站在系统中重要程度比不上大型发电厂和枢纽变电站，没必要追求可靠性过高的复杂接线形式，与系统的接入方式可以采用单回路弱联系方式接入。然而部分中、小型发电厂、变电站靠近负荷中心，对于有进去负荷的情况，宜采用可靠性较高的母线接线形式，以便适应近区各类负荷对供电可靠性的要求。

2. 负荷性质和类别

负荷按其重要性有Ⅰ类、Ⅱ类、Ⅲ类三种，主接线的设计应当与负荷类别相适应。

Ⅰ类负荷：中断供电将造成人身伤亡；中断供电将在政治、经济上造成重大损失（如重大设备损坏、重大产品报废、重要原料生产的产品大量报废、国民经济中重点企业连续生产过程被打乱需要长时间恢复等）；中断供电将影响有重大政治、经济意义的用电单位工作（例如重要交通枢纽、重要通信枢纽、大型体育场馆等）；中断供电将发生中毒、爆炸、火灾等情况的负荷，以及特别重要场所不允许中断供电的负荷。

Ⅱ类负荷：中断供电将在政治、经济上造成较大损失时（例如主要设备损坏、大量产品报废、连续生产过程被打乱需较长时间才能恢复、重点企业大量减产等）；中断供电将影响重要用电单位的正常工作（例如交通枢纽、通信枢纽等用电单位中的重要电力负荷，以及中断供电将造成大型影剧院、大型商场等较多人员集中的重要的公共场所秩序混乱）。

不属于Ⅰ类、Ⅱ类负荷的其他负荷，属于Ⅲ类负荷。

根据《电力工程电气设计手册》要求，Ⅰ类负荷必须有两个独立电源供电，当任何一个电源失去后，能保证全部Ⅰ类负荷不间断断电。Ⅱ类负荷一般要有两个独立电源供电，当任何一个电源失去后，能保证全部或者大部分Ⅱ类负荷的供电。对于Ⅲ类负荷一般需要一个电源供电。

因此，对于有重要地位的Ⅰ类、Ⅱ类负荷，主接线必须采用可靠性较高的形式，且保证两路电源供电。

3. 设备制造水平

主接线的可靠性除了取决于接线方式，还在很大程度上取决于设备本身的可靠性，采用可靠性更高的设备可以一定程度上简化接线。

大容量机组、自动装置和先进技术的应用，都有利于提高主接线的可靠性，但是并不是设备越多越新越复杂就越可靠；相反，增加一些不必要的设备会使得接线更加复杂，运行不便，导致主接线可靠性降低。

4. 长期运行经验

电网的可靠运行，不仅取决于"物"的因素，还取决于"人"的因素。可靠性与运行管理水平和运行值班人员的素质有密切关系，衡量可靠性的客观标准是运行实践。国内网长期运行经验的累积，经过总结均反映于相关的技术规程、规范之中，在主接线设计时，应当遵循。

主接线的可靠性通常从以下几个具体的方面来衡量：断路器检修时，是否影响对系统的供电；线路、断路器或者母线故障时，以及母线侧隔离开关检修时，停运的出线回路数和停电时间，并且是否能够保障全部Ⅰ类和大部分Ⅱ类用户的供电；避免发电厂、变电站全站停电的可能性；大型机组停运时，是否危及系统的稳定运行。

在可靠性分析中，最主要的基础统计数据是断路器的可靠性，主要指标是故障率、可用系数和平均修理小时数。评估供电可靠性的主要指标是停电频率、每次停电持续时间以及停电时产生损失或代价。

二、灵活性

主接线的灵活性是指同一种电气主接线有不同的运行状态，并且能够灵活地在各种运行方式之间转换。灵活性包括以下几个方面。

1. 操作的方便性

电气主接线应该在服从可靠性的基本要求下，接线简单，操作方便，尽可能地减少操作步骤和复杂程度，以便运行人员掌握，降低误操作可能性。

2. 调度的方便性

电气主接线在正常运行时，要能够根据调度要求方便地改变运行方式，并且在发生事故时，能够尽快切除故障，使停电时间最短、影响最小。

3. 扩建的方便性

对将来需要扩建的发电厂、变电站，其主接线设计必须要有扩建方便性。特别是对于很多火电厂和变电站，在设计时应当留有发展的余地。所以在设计时应当考虑三个阶段，首先是最初的接线方式的完整性；其次是最终扩建后接线方式的完整性；最后是从最初接线过渡到最终接线的施工的可行性，并且扩建要尽可能不影响原有接线的连续供电，或者降低扩建对其的影响。

三、经济性

在设计主接线时，最大的矛盾集中在可靠性与经济性上，因为要求可靠性越高，成本也就越高，经济性就越差。所以在设计主接线时，应该在刚好满足可靠性和灵活性的前提下，做到经济合理。经济性主要从以下几个方面考虑。

1. 节省一次性投资

主接线设计应当简单清晰，并且采取适当的方式限制最大短路电流，则在主接线设计后选择电气设备时，可以选用成本更低的电气设备，以节约一次性投资。对于大容量发电厂和变电站，在条件允许的情况下，尽量采用分期投资建设，则可以尽快地发挥经济效益。

2. 减少占地面积

主接线设计需要为配电装置的布置规划占地，应当在满足安全和可靠性、灵活性的前提下，缩小配电装置所占土地面积。

3. 降低电能损耗

在发电厂或变电站中，电能损耗主要来自变压器，应按照设计要求合理地选择变压器的型式、容量、台数和厂家。

Program ③ Design of Main Electrical Wiring

As the primary part of the power plant and substation design, the main electrical wiring is an important part of power system. The basic forms and characteristics of the main electrical wiring have been introduced in Program 1 of this book. Centered on the design of main electrical wiring, the basic requirements, procedures and steps of design for the main electrical wiring of substation, power plant and power station are introduced, as well as the main transformer selection and the methods of limiting short-circuit current are analyzed in this program.

Module 1   Basic Requirements of Design

The main electrical wiring is in the HV electrical major structure and heavy current of the power plant and substation and an important part of power system. The main wiring is determined in close relation to the reliability, flexibility, safety and economical efficiency of the overall operation of the power system, with decisive influence on the relay protection of power system, the selection of automatic devices and equipment, and the PDU layout at the same time. Therefore, the main wiring must be designed in comprehensive consideration of various factors, and fully demonstrated and compared to finally obtain the best solution in line with the actual situation.

The basic requirements for the design of main electrical wiring include reliability, flexibility and economical efficiency.

1. Reliability

The primary task of power generation is to guarantee the safety and reliability of power supply, which is the most basic requirement for main electrical wiring. Power outage will not only cause losses to power plants, but also lead to economic losses to the society. In the regions with developed economy, power outage caused by faults may cause equipment damage, product scrapping, railway system breakdown, city life chaos, and even personal injuries, resulting in huge economic losses, social issues and safety issues. Therefore, the form of main wiring must be able to guarantee the power supply reliability.

Main wiring reliability is not absolute. The main wiring in the same form that is reliable for some power plants and substations may not be reliable for some other power plants and substations. Therefore, when analyzing the main wiring reliability, it is required to fully consider the functions and status of the power plant and substation in the system, as well as many factors, such as nature and category of user load, and manufacturing level and operating experience of equipment.

1) Functions and Status of Power Plant and Substation in the Power System

The main electrical wiring of each power plant and substation shall be fit in the power system connected.

Due to large power supply volume, wide range and important position of the large power plant or load-center substation in the power system, any accident may damage the steady operation of the entire system, causing massive losses. Therefore, the forms of main wiring with quite high reliability shall be adopted for large power plants or load-center substations. Generally, large power plants and load-center substations are far from the load center, in which there are high voltage and large capacity. Therefore, strong connection forms should be adopted for system access, such as double circuit or looped network, guaranteeing that the wiring method at corresponding voltage class is reliable. Qualitative analysis and quantitative calculation are necessary for the main wiring reliability at the time of design.

Small and medium-sized power plants and substations are less important in the system than large power plants and load-center substations, and it is unnecessary to seek after complex wiring forms with too high reliability. The weak connection, such as single circuit, can be used for system access. However, part of small and medium-sized power plants and substations are near the load center. In case of incoming load, the bus connection forms with high reliability are advisable, facilitating the conformance to the power supply reliability requirements of various loads in the nearby areas.

2) Load Nature and Category

Loads are divided into three types by importance: Class I, Class II and Class III. The main wiring shall be design to the extent fit in the load category.

Class I load: Interruption of power supply will cause personal injuries. Interruption of power supply will cause heavy political and economic losses (such as damage to major equipment, scrapping of major products, scrapping of a large number of products produced by important raw materials, and need for long-term recovery after the continuous production process of key enterprises in the national economy is disrupted). Interruption of power supply will affect the work of the units consuming electricity with significant political and economic significance (such as important transportation junctions, important communication centers and large-scale stadiums). Interruption of power supply will generate the loads in such situations as poisoning, explosions and fires, as well as the loads not allowed to be interrupted in particularly important places.

Class II load: Interruption of power supply will cause relatively heavy political and economic losses (such as main equipment damage, scrapping of a large number of products, need for long-term recovery after the continuous production process is disrupted, and reduction of output of key enterprises drastically). Interruption of power supply will affect the normal work of important units consuming electricity (such as critical power loads of such units consuming electricity as transportation junctions and important communication centers), and it will result in chaos in important public places with a large of number of people concentrated, such as large cinemas and shopping malls).

The loads not classified as Class I load and Class II load are called Class III load.

According to the *Electrical Design Manual for Power Engineering*, the Class I load must be powered on by two independent power supplies. Uninterrupted outage of Class I load shall be guaranteed completely after either power is lost. Generally, the Class II load should be powered on by two independent power supplies. Powering on the Class II load completely or mostly shall be guaranteed after either power is lost. In general, the Class III load should be powered on by one power supply.

Therefore, the main wiring with relatively high reliability must be adopted for important Class I load and Class II load, and the two-way power supply shall be guaranteed.

3) Equipment Manufacturing Level

The reliability of main wiring depends on the wiring method, and also on the reliability of the equipment itself to a great extent. The wiring method can be simplified to some extent after the equipment with higher reliability is adopted.

Application of large-capacity units, automatic devices and advanced technologies facilitates the reliability enhancing of main wiring. However, it does not mean that the reliability is higher if there is more equipment, or newer equipment or more complex equipment. On the contrary, additional unnecessary equipment will make the wiring more complex and the operation inconvenient, leading to reduced reliability of main wiring.

4) Long-term Operating Experience

Power grids run reliably depending not only on the "objective" factor, but also on the "human" factor. Reliability is closely bound up with the operation and management level and the quality of operators on duty, and operation practices are objective standards to evaluate reliability. The accumulative long-term operating experience in domestic grids is manifested in relevant technical regulations and specifications after being summed up, which shall be followed at the time of main wiring design.

The reliability of main wiring is usually evaluated from the followings specifically: The possibility of influencing the power supply to the system by the maintenance of circuit breaker; the number of outgoing circuits stopping the service and the duration of power outage, as well as the ability to ensure the power supply to all Class I and most Class II users when the lines, circuit breakers or buses fail and during the maintenance of disconnector at the bus side; avoidance of the possibility that power plants and substations have power outage completely; and the possibility of jeopardizing the steady system operation when large units stop the service.

In terms of reliability analysis, the reliability of circuit breakers is regarded as the uppermost basic statistical data, and main indicators include failure rate, availability coefficient and mean repair hours. Main indicators for evaluating the power supply reliability include outage frequency, duration of each outage, and losses or costs from outage.

2. Flexibility

The flexibility of main wiring refers to that the same main electrical wiring is in different operating states with the ability of flexible switching between operating modes. Flexibility is shown as follows.

1) Operation Convenience

The main electrical wiring shall be capable of simple wiring and easy operation with operation steps and complexity minimized while meeting the basic requirements of reliability, so that operators will grasp the wiring and reduce the possibility of misoperation.

2) Scheduling Convenience

During normal operation, the main electrical wiring shall be able to easily change the mode of operation as required by scheduling, and to eliminate faults as soon as possible in accidents to minimize the duration of power outage and the impact.

3) Expansion Convenience

Expansion convenience is a must for the power plants and substations to be expanded in the future during the main wiring design. In particular, room for expansion shall be left in design for many thermal power plants and substations. Therefore, three stages shall be considered during design: first, the integrity of the initial wiring method; second, the integrity of wiring method after final expansion; and finally, the feasibility of construction from the initial wiring to the final wiring. In addition, expansion shall not affect the continuous power supply of the original wiring to the greatest extent nor reduce the impact on it.

3. Economical Efficiency

When designing the main wiring, the biggest contradiction exists in reliability and economical efficiency. This is because the cost will be higher and the economical efficiency will be worse if the required reliability is higher. Thus, when designing the main wiring, economic rationality is required while just meeting the requirements of reliability and flexibility. Economical efficiency shall be considered from the followings mainly.

1) One-off Investment Reduction

The main wiring shall be designed in a simple and clear way, and the maximum short-circuit current shall be limited by appropriate means. Thus, when selecting electrical equipment after main wiring design, the electrical equipment with lower cost can be selected, saving the one-off investment. Large-capacity power plants and substations shall be constructed based on staged investment to the greatest extent if conditions permitted, achieving economic benefits as early as possible.

2) **Floor Area Reduction**

During the main wiring design, the land shall be planned for the layout of PDUs, and the land area for PDUs shall be reduced on the premises of safety, reliability and flexibility.

3) **Electric Energy Loss Reduction**

Electric energy loss of power plants or substations is mainly from transformers, and the type, capacity, number and manufacturer of transformers shall be reasonably selected as per design requirements.

模块二　设计程序和步骤

一、发电厂设计的基本程序

发电厂和变电站基本设计程序需要按照我国的基本程序进行，主要有设计前期工作阶段、设计工作阶段、施工运行阶段三个阶段。下文以发电厂设计为例，具体步骤如表 3-1 所示。

表 3-1　设计建设基本程序

设计阶段	设计基本程序	任务
设计前期工作阶段	初步可行性研究	对建厂地区进行地区调查，进行比较论证，推荐可能的厂址、规模和建设顺序，为编制和审批项目建议书提供依据
	编制项目建议书	提出建厂的必要性和负荷、建厂性质和规模、建厂厂址和条件、建厂年份和顺序、投资控制和筹措等
	可行性研究	落实建厂条件，确定建厂规模，提出设计原方案，完成环境影响报告书，进行全面综合性技术经济分析论证和方案比较，提出投资估算和经济效益评价，取得外部条件的协议书，为编制和审批设计任务书提供可靠依据
	编制设计任务书	明确建设目的、依据、建设规模、建设条件、主要协作配合条件、主要工艺流程、环境保护要求、建设地点、占地面积、建设进度控制、投资和劳动定员控制、需要研制的新产品等
设计工作阶段	初步设计	确定建设标准、各项技术原则和总概算，以便编制投资计划、实行投资包干、控制工程拨款、组织主要设备订货、进行施工准备，作为施工图设计依据
	施工图设计	为订货、施工运行的依据，经审定的预算为预算包干、工程结算的依据
施工运行阶段	配合施工	交代设计意图，解释设计文件，及时解决工程管理与施工中设计方面出现的问题，参加试运转，参加竣工验收和投产
	运行回访或总结反馈	总结和积累设计上的经验教训，编入总结报告以改进设计、提高水平

二、发电厂设计的具体步骤

电气主接线设计属于发电厂、变电站设计和建设中的内容,发电厂、变电站设计的要求、任务不同,其设计的深度、广度也有所差异,但是主接线的设计原则、方法和步骤基本相同。下文以发电厂设计为例介绍具体步骤。

(一)对原始资料的分析

1. 工程情况

工程情况包括发电厂类型、设计规划容量、单机容量及台数、最大负荷利用小时数及可能的运行方式等。

发电厂容量的确定与国家或地区经济发展计划、电力负荷增长速度、系统规模、电网结构以及备用容量等因素有关。发电厂装机容量标志着发电厂的规模和在电网中的地位和作用。在设计时,可优先选用大型机组,但是最大单机容量不大于系统总容量的10%,以保证该机组在检修或者事故时,系统的供电可靠性。

发电厂运行方式和年利用小时数直接影响主接线的设计。不同类型的发电厂工作特性是不一样的,运行方式和年利用小时数也各不相同,相应对主接线要求也不相同。比如,承担基荷的大型核电厂、火电厂或水电厂等主接线以较高的可靠性作为基本要求进行选择,承担调峰、调相任务的灵活的中、小型水电厂,只需要根据其水能利用和库容等状态承担一定的基荷、腰荷,所以在设计主接线时,应当以较高的灵活性为基本要求。

2. 电力系统的情况

电力系统的情况包括电力系统近期、远期发展规划,发电厂在系统中的地位和作用,本期工程中近期和远期与电力系统连接的方式,以及各级电压中性点接地方式等。

如果发电厂的总容量在当地电力系统中占比较大,电厂处于重要地位,则应选择可靠性较高的主接线形式,因为此类电厂一旦全厂停电,会影响到整个系统供电可靠性。

为简化网络结构以及电厂主接线,减少电压等级,电厂接入的系统电压等级不应超过两级,且出线应当尽量减少,利于简化配电装置的规模及维护。

变压器或发电机中性点接地方式是一个综合性问题,它与电压等级、单相接地短路电流、过电压水平、继电保护配置等均有关系,直接影响电网的绝缘水平、系统供电可靠性、主变压器和发电机的运行安全以及对通信线路的干扰等。我国一般对 35 kV 及以下变压器采用中性点非直接接地方式,也称为小电流接地系统;对 110 kV 及以上变压器采用中性点直接接地系统,又称为大电流接地系统;发电机中性点都采用非直接接地,目前广泛使用中性点经消弧线圈或经单相配电变压器接地。

3. 负荷情况

负荷情况包括负荷的性质、地理位置、输电电压等级、出线回数和输送容量等。

电力系统原始资料是设计主接线的基础数据。电力负荷预测是电力规划工作的重要组成部分,也是电力规划的基础。对电力负荷的预测不仅应有短期负荷预测,还应有中长期负荷预测,其预测准确性直接关系到主接线设计成果的质量。

发电厂承担的负荷应尽可能使全部机组安全满发,并按系统提出的运行方式,在机组之间经济合理分布负荷,减少母线上电流,降低厂内线损,使电机运转稳定和保证电能质量。

4. 环境条件

环境条件包括当地气温、湿度、覆冰、污秽、风向、水文、地质、海拔以及地震等因素。外界环境对电气主接线设计以及设备选择均有影响,在设计时亦需要充分考虑。

5. 设备供货情况

设备供货情况往往是设计能否成立的重要前提,为使主接线具有可行性,必须对各主要电气设备的性能、制造力和供货情况、价格等因素进行汇集并比较分析。

(二)主接线方案拟定与选择

在原始资料分析的基础上,根据对电源、出线回路数、电压等级、变压器类型、台数、容量以及母线结构等因素的考虑,可以拟定若干个主接线方案。

根据发电厂对主接线的基本要求,从技术上论证并淘汰一些不满足基本要求的不合理方案,最终保留几个技术相当、均能够满足任务要求的方案,再进行经济性比较。对于在系统中占有重要地位的发电厂,还应进行可靠性的定性、定量分析计算比较,最终确定出在技术上合理、经济上可行的最终方案。拟定主接线方案的具体步骤如下:

(1)根据发电厂和电网的具体情况,初步拟定出若干技术基本可行的接线方案。

(2)选择主变压器台数、容量、形式、参数及运行方式。

(3)拟定各电压等级的基本接线形式。

(4)确定自用电的接入点、电压等级、供电方式等。

(5)对上述各部分进行合理组合,拟定出数个初步可靠方案,然后从可靠性、灵活性等各方面分析比较,确定2~3个待选方案。

(6)对待选方案进行经济性比较,最终确定主接线方案。

其中技术性分析比较除了可靠性、灵活性,还包括接入系统后稳定性、电能质量、安全性、自动化程度、需要研制的新设备、新技术的应用、扩建等方面。

(三)短路电流计算和主要电气设备选择

按照不同电压等级各类电气设备的选择与校验的要求,确定电气主接线各短路计算点,进行短路电流计算,合理选择除主变压器外的其他电气设备。

(四)绘制电气主接线图

将最终确定的电气主接线按工程要求绘制施工图。

(五)编制工程概算

对于工程设计,无论哪个设计阶段,概算都是必不可少的部分。它不仅反映工程设计的经济性与可靠性关系,而且为合理确定和有效控制工程造价创造条件,为工程付诸实施,为投资包干、招标承包等正确处理经济利益提供基础。概算的编制是以设计图纸为基础,以国家颁布的相关规定为依据,按照国家定价,结合市场浮动等原则进行。

Module 2　Design Procedures and Steps

1. Basic Procedures of Power Plant Design

The basic design procedures of power plants and substations shall be subject to the basic procedures in China, mainly including three stages, i.e. pre-design stage, design stage, and construction and operation stage. Power plant design is taken as an example in this section, and the specific steps are shown in Table 3-1.

Table 3-1　Basic Procedures of Design and Construction

Design stage	Basic procedure of design	Tasks
Pre-design stage	Prefeasibility study	Survey the areas with plants built, perform the comparative argument, and recommend possible plant site, scale and construction sequence, providing the basis for preparing and approving the project proposal.
	Preparation of project proposal	Put forward the necessity and load, nature and scale, location and conditions, year and sequence, investment control and fundraising of plant construction.
	Feasibility study	Determine the conditions and scale of plant construction, propose the original design scheme, complete the environmental impact statement, perform the overall comprehensive technical and economic analysis and compare schemes, put forward the investment estimation and economic benefit evaluation, and obtain the agreements on external conditions, providing the reliable basis for preparing and approving the design specification.
	Preparation of design specification	Specify the purpose, basis, scale and conditions of construction, main collaboration and cooperation conditions, main process flows, environmental protection requirements, location of construction site, floor area, construction progress control, control over investment and manpower quota, and new products to be developed.
Design stage	Preliminary design	Determine the construction standards, technical principles and general overview, facilitating investment plan preparation, implementation of investment lump-sum contracting, project appropriation control, organization for ordering of main equipment, and construction preparation, which are regarded as the basis for construction drawing design.
	Construction drawing design	Provide the basis for ordering, construction and operation. The approved budget is the basis for lump-sum budget and project settlement.
Construction and operation stage	Cooperation in construction	Account for the design intent, explain the design documents, promptly solve the problems arising from the design of project management and construction, participate in the test run, participate in the completion acceptance and put into production.
	Return visit to operation or summary and feedback	Sum up and accumulate the design experience and draw lessons, and include it in the final report to improve the design and level.

2. Specific Steps of Power Plant Design

The main electrical wiring design is included in the design and construction of power plants and substations. As the design requirements and tasks are different for power plants and substations, differences also exist in the depth and breadth of design. However, design principles, methods, and steps of main wiring are basically the same. Power plant design is taken as an example in this section, and the specific steps are shown below.

1) Analysis of Original Data

(1) Situation of the project.

Situation of the project include type, designed and planned capacities, unit capacity and number, maximum load utilization hours, and possible operating modes of power plant.

Determination of power plant capacity is related to such factors as national or regional economic development plans, power load growth rate, system scale, power grid structure and reserve capacity. The installed capacity of power plant marks its scale and its position and functions in power grid. At the time of design, large units are preferred, but the maximum unit capacity cannot exceed 10% of the total system capacity, thus guaranteeing the power supply reliability of the system during the unit maintenance or accidents.

The operation mode and annual utilization hours of the power plant will directly affect the main wiring design. Different types of power plants have different working characteristics, varied operation modes and annual utilization hours and diversified main wiring requirements accordingly. For example, relatively high reliability shall be regarded as the basic requirement to select the main wiring of large nuclear power plants, thermal power plants or hydropower plants that bear the base load, while relatively high flexibility shall be regarded as the basic requirement to design the main wiring of the flexible medium-sized and small hydropower plants used for peak shaving and phase modulation which will only bear certain base load and shoulder load based on their hydropower utilization and storage capacity.

(2) Condition of power system.

Condition of power system include the short-term and long-term development plans of power system, position and functions of power plants in the system, methods of connection to the power system in the near term and long term of the project in this phase, and the grounding modes of neutral points at all voltages.

If the total capacity of power plant occupies a relatively large proportion in the local power system and the power plant is on the front burner, the highly reliable main wiring form shall be selected. This is because the power supply reliability of the entire system will be affected once such type of power plant has power outage.

In order to simplify the network structure and the main wiring of power plant, and to lower the voltage class, at most two voltage classes are required for the system connected by the power plant. In addition, outgoing lines shall be minimized, facilitating the simplification of PDU scale and maintenance.

As a complex matter, the neutral point grounding mode of transformer or generator is related to voltage class, single-phase grounding short-circuit current, overvoltage level and relay protection configuration, which directly affects the insulation level of power grids, the power supply reliability of system, the operation safety of main transformer and generator, and the interference to communication line. In China, the neutral-point indirect grounding mode is generally adopted for the transformers of 35 kV and below, which is also known as low current grounding system. The neutral point direct grounding mode is adopted for the transformers of 110 kV and above, which is also known as heavy current grounding system. Neutral points of generators are all indirectly grounded. At present, neutral points are widely grounded through arc suppression coil or single-phase distribution transformer.

(3) Load condition.

Load condition include load nature, geographical location, transmission voltage class, number of outgoing lines, and transmission capacity.

The original data of power system is regarded as the basic data of main wiring design, and the power load forecasting is an important part and foundation of power planning. Power load forecasting includes short-term and mid-and-long term load forecasting, and their accuracy will directly affect the quality of main wiring design achievements.

The loads borne by the power plant shall allow that all units generate the power safely to the greatest extent, and shall economically and reasonably distribute the loads between units based on the operation mode raised in the system, reducing the current on the bus and the line loss in the plant, so that the motor will run stably with the power quality guaranteed.

(4) Environmental conditions.

Environmental condition include local temperature, humidity, icing, fouling, wind direction, hydrology, geology, altitude, and earthquakes. External environment impacts the design of main electrical wiring and the equipment selection, which shall be fully considered during design.

(5) Equipment supply situation.

The situation of equipment supply is often an important precondition for the feasibility of design. In order to make the main wiring feasible, the performance, manufacturing capacity, supply, price and other factors of main electrical equipment shall be gathered and compared.

2) Formulation and Selection of Main Wiring Scheme

Based on original data analysis, several main wiring schemes can be formulated in consideration of such factors as power supply, number of outgoing circuits, voltage class, type, number and capacity of transformers, and bus structure. Some unreasonable schemes not in line with basic requirements are technically demonstrated and eliminated based on the basic requirements of the power plant for main wiring. Several schemes technically equivalent and able to meet the task requirements are retained finally for economic comparison. For the power plants playing an important role in the system, qualitative and quantitative analysis, calculation and comparison of reliability are required to ultimately determine the final scheme technically reasonable and economically feasible. The specific steps for the formulation of main wiring scheme are as follows:

(1) Preliminarily draft several wiring schemes which are basically feasible in technology based on the specific situations of power plant and power grid.

(2) Select the number, capacity, form, parameters and operation mode of main transformers.

(3) Formulate the basic wiring forms at all voltage classes.

(4) Determine the access points, voltage class and power supply methods of auxiliary power supply.

(5) Reasonably combine those mentioned above, formulate several preliminarily reliable schemes, and then perform analysis and comparison in terms of reliability and flexibility, and identify two or three alternatives.

(6) Compare the alternatives in terms of economical efficiency, to ultimately determine the main wiring scheme.

Technical analysis and comparison include reliability and flexibility, as well as stability, power quality, safety, automation degree, new equipment to be developed, and application and expansion of new technologies after access to system.

3) Short-circuit Current Calculation and Main Electrical Equipment Selection

As required by the selection and validation of various electrical equipment at different voltage classes, all short-circuit calculation points of main electrical wiring shall be determined, the short-circuit current shall be calculated, and other electrical equipment except for main transformer shall be reasonably selected.

4) Drawing of Main Electrical Wiring Diagram

Construction drawings shall be drawn based on the finally determined main electrical wiring as required by the project.

5) Preparation of Project Budget

The budget is essential in all engineering design stages. It not only manifests the relationship between the economical efficiency and reliability of engineering design, but also creates conditions for reasonable determination and effective control of project cost, provides the basis for the project implementation and the correct handling of economic benefits from investment lump-sum contracting, invitation for bids and contracting. The budget shall be prepared based on design drawings, relevant regulations promulgated by the state and the state fixed prices on the principle of combining market floats.

模块三　主变压器的选择

在发电厂和变电站中,用来向电力系统或用户输送功率的变压器称为主变压器;用于两种电压等级之间交换功率的变压器称为联络变压器;只提供自用电的变压器称为自用变压器或厂用变压器、站用变压器。

主变压器属于发电厂、变电站重要的一次设备,但根据发电厂设计的具体步骤可知,该设备应当在主接线设计之前做选择,而非短路计算完成后进行选择。

一、主变压器容量和台数

主变压器的容量、台数直接影响主接线的形式和配电装置的结构。主变压器的选择依据除了输送容量外,还应根据电力系统未来 5~10 年发展规划、输送功率大小、馈线回路数、电压等级以及接入系统的紧密程度等因素,进行综合分析和合理选择。如果主变压器容量选择得过大,或台数过多,则增加了一次性投资,增大了占地面积,增加了运行电能损耗,设备也未能充分发挥效用,直接导致主接线整体经济性较差;如果主变压器容量过小,则会使得无法满足负荷需求,或发电机剩余功率无法输出给用户,造成电力资源浪费,这在技术上是不合理的。因为每千瓦时的发电设备投资远比每千瓦时的变电设备投资更大。因此,在选择发电厂的主变压器时,应遵循以下基本原则。

(一) 单元接线的主变压器

单元接线的主变压器应当按下列条件中的较大者选择:
(1) 发电机的额定容量扣除本机组的厂用电负荷后,再留 10% 的裕度。
(2) 发电机的最大连续容量,扣除本机组的厂用负荷。

采用扩大单元接线时应采用分裂绕组变压器,其容量也应当按上述原则进行计算,算出的两台容量之和即分裂变压器的容量。

(二) 具有发电机电压母线接线的主变压器

连接在发电机电压母线与系统之间的主变压器容量应考虑以下因素:
(1) 当发电机全部投入运行时,发电机母线上负荷最小时,主变压器应能将发电机电压母线上的剩余有功和无功容量送入系统。
(2) 当接在发电机电压母线上的最大一台机组故障或检修时,或者其他原因需要限制本厂输出功率时,主变压器应能从系统倒送功率,保证发电机电压母线上最大负荷的需要。
(3) 如果发电机电压母线接有 2 台及以上的主变压器时,当其中一台容量最大的主变压器因检修或故障退出运行时,其他主变压器应当能够输送母线剩余功率的 70% 以上。
(4) 在电力市场环境下,中、小型火电厂在热力负荷需求减少等情况下,可能会停用火电厂部分或全部机组,主变压器应具有从系统中倒送功率的能力,满足发电机电压母线上最大负荷需求。

（5）为确保对发电机电压上的负荷供电可靠性，介于发电机电压母线上的主变压器不应少于 2 台，其总容量除满足以上几点要求外，还应考虑不少于 5 年负荷的发展。

（6）对于利用工业产生的余热发电的中、小型电厂，可以只装 1 台主变压器与电力系统构成弱连接。

（三）连接两种升高电压母线的联络变压器

联络变压器的台数一般只设置 1 台，最多不超过 2 台，容量选择应考虑以下两点：

（1）联络变压器容量应满足两种电压的网络在各种不同运行方式下有功功率和无功功率的交换。

（2）联络变压器容量一般不应小于接在两种电压母线上的最大一台机组容量，以保证最大一台机组故障或检修时，通过联络变压器来满足故障、检修侧的负荷要求；同时也可在线路检修或故障时，通过联络变压器将剩余容量输送入另一系统。

（四）变电站变压器

（1）变电站主变压器容量应按 5~10 年规划负荷来选择，并适当考虑远期 10~20 年负荷发展，对于城郊变电站，主变压器容量应与城市规划相结合。

（2）根据变电站所带负荷的性质和电网结构来确定主变压器的容量。

（3）对于有重要负荷的变电站，应考虑一台主变压器停运时，其余变压器容量在计及过负荷能力后的允许时间内，应保证用户的Ⅰ类和Ⅱ类负荷。

（4）对于一般性变电站，当一台主变压器停运时，其余变压器容量应能保证全部负荷的 70%~80%。

（5）对大城市郊区的变电站，在中、低压侧已构成环网的情况，变电站以装设 2 台主变压器为宜；对地区性孤立的变电站或大型工业专用变电站，在设计时应考虑 3 台主变压器的可能性；对于规划只装设 2 台主变压器的变电站，其变压器基础宜按大于变压器容量设计，以便应对负荷的发展。

二、变压器形式和结构

（一）相　数

主变压器采用三相或者单相，主要考虑变压器的制造条件、可靠性要求以及运输条件等因素。特别是大型变压器，尤其是要考虑运输的可能性，以保证运输尺寸不超过隧洞、涵洞、桥洞等允许通过的限额，运输重量不能超过路面、桥梁、车辆、船舶等允许承载的能力。

单相变压器与三相变压器相比，投资大、占地面积大、运行损耗大、配电装置结构复杂、维修工作量大，但是 3 台单相变压器的运输比 1 台同容量三相变压器更加灵活，要求更低。

所以对于主变压器的相数，应考虑以下原则：

（1）当不受运输条件限制时，330 kV 及以下均应使用三相变压器。

（2）对于 500 kV 变压器，应进行技术、经济比较后，确定选用三相变压器、2 台半容量三相变压器或者 3 台单相变压器。

（二）绕组数

电力变压器按其每相的绕组数可分为双绕组、三绕组或更多绕组；按电磁结构可分为普通双绕组、三绕组、自耦式以及低压绕组分裂式等。

发电厂以两种升高电压向用户供电时，可以采用 2 台双绕组变压器或者三绕组变压器。

机组容量为 125 MW 及以下的发电厂多采用三绕组变压器，但是三绕组变压器每个绕组通过容量应达到额定容量的 15% 以上，否则绕组未能充分利用，不如采用 2 台双绕组变压器更经济。此外，发电厂或变电站采用的三绕组变压器不宜多于 3 台，可以避免中压侧布置的复杂性，降低成本及中压侧短路容量。

机组容量为 200 MW 以上的发电厂通常采用"发电机-双绕组变压器"单元接入系统，两种升高电压级之间采用联络变压器。

联络变压器一般选择三绕组变压器，其低压绕组可接高压厂用备用变压器或无功补偿装置。

扩大单元接线主变压器，应优先选用低压分裂绕组变压器，可限制短路电流。

对 220 kV 及以上有三个电压等级的变压器，可优先选用自耦变压器，其特点为损耗小、价格低、效率高、阻抗小。

（三）调压方式

为保证电网的电能质量，电压必须维持在允许范围内。通过改变变压器分接头开关，可以改变变压器高压绕组匝数，从而改变其变比，实现一定范围内电压的调整。但是这种方法无法应对电网无功容量不足导致的电网电压下降。

变压器分接头切换形式有两种：一种是不带负荷切换，称为无载调压或者无励磁调压，结构较为简单，但只能停电切换；另一种是可以带负荷切换，称为有载调压，可调节范围达到 30%，结构复杂，价格昂贵，但是可以在不停电的情况下完成调压，在以下情况下予以使用：

（1）接于输出功率变化大的发电厂的主变压器，特别是潮流方向不固定，且要求变压器二次电压维持在一定水平时。

（2）接于时而为送电端、时而为受电端、具有可逆工作特点的联络变压器，为保证供电质量，要求母线电压恒定时。

通常发电厂因为有发电机存在，可以通过调节励磁来调节电压，所以其主变压器很少采用有载调压。

（四）冷却方式

油浸式变压器冷却方式随其电压等级、型式、容量的不同而不同，一般有自然风冷却、强迫风冷却、强迫油循环风冷却、强迫油循环水冷却、强迫油循环导向冷却等型式。

容量在 31.5 MV·A 以下的中、小型变压器通常采用变压器油箱上的片状或者管型辐射式冷却及电动风扇的自然风冷却以及强迫风冷却等方式。

容量在 31.5 MV·A 及以上的变压器，一般采用强迫油循环风冷却，在发电厂水源充足的情况下，也可采用强迫油循环水冷却。

容量在 350 MV·A 及以上的特大型变压器一般采用强迫油循环导向冷却。

Module 3　Selection of Main Transformer

In power plants and substations, the transformer used to transmit the power to the power system or users is known as main transformer. The transformer used to exchange the power between two voltage classes is known as interconnecting transformer. The transformer only providing the auxiliary power supply is known as self-use transformer, plant transformer or station transformer.

As the important primary equipment of power plants and substations, the main transformer shall be selected before main wiring design rather than after the short-circuit calculation from the specific steps of power plant design.

1. Capacity and Number of Main Transformers

The capacity and number of main transformers will directly affect the main wiring form and the PDU structure. The main transformer shall be selected based on the transmission capacity, and also comprehensively analyzed and reasonably selected based on such factors as development plan of the power system in the next 5—10 years, size of transmission power, number of feeder circuits, voltage class, and tightness of access into system. Excessive capacity selected for the main transformer or too many main transformers (if any) will increase the one-off investment, floor area and electric energy loss during operation, and cause the failure of equipment operation to be effective fully, leading to poor overall economic performance of main wiring. Too small capacity of the main transformer (if any) will result in the failure to meet the load demands, or to output the surplus power of generator to users, causing the waste of power resources, which is technically unreasonable. This is because the investment in the power generation equipment per kilowatt hour is more than that in substation equipment per kilowatt hour. Therefore, main transformers of power plants shall be selected subject to the following basic principles.

1) Main Transformer with Unit Wiring

The main transformer with unit wiring shall be selected based on the larger one among the following conditions:

(1) After the auxiliary power load of the unit is deducted from the rated capacity of the

generator, 10% margin shall be reserved.

(2) The auxiliary power load of the unit shall be deducted for the maximum continuous capacity of generator.

For the expanded unit wiring, the split winding transformer shall be used, and its capacity shall also be calculated on the above principles. The capacity of the split transformer equals to the sum of two calculated capacities.

2) Main Transformer with Generator Voltage Bus Connection

The following factors shall be considered for the capacity of the main transformer connected between the generator voltage bus and the system:

(1) When all generators are put into operation with the lowest load on the generator bus, the main transformer shall be able to transmit the remaining active and reactive capacities of the generator voltage bus into the system.

(2) If the output power of the plant should be limited due to the faults or maintenance of the largest unit connected to the generator voltage bus, or for other reasons, the main transformer shall be able to reversely transmit the power from the system, guaranteeing the demand for the maximum load on the generator voltage bus.

(3) If the generator voltage bus is connected to two or more main transformers, when one of the main transformers with the largest capacity quits running due to maintenance or faults, other main transformers shall be able to transmit over 70% of the surplus power of the bus.

(4) In the electricity market environment, part of or all units of the thermal power plant may be shut down due to the demand reduction of medium-sized and small thermal power plants in heat load, and the main transformer shall be able to reversely transmit the power from the system, guaranteeing the demand for the maximum load on the generator voltage bus.

(5) In order to ensure the reliability of load-based power supply on the generator voltage, at least two main transformers are required on the generator voltage bus. Their total capacity shall meet the above requirements with the consideration of the load development for at least five years.

(6) For the small and medium-sized power plants that utilize industrial waste heat to generate power, only one main transformer can be installed to weakly connect the power system.

3) Interconnecting Transformer Connecting Two Boosted Voltage Buses

Generally, only one interconnecting transformer is installed (at most two interconnecting transformers are allowed). The capacity shall be selected in consideration of the following two aspects:

(1) The capacity of interconnecting transformer shall meet the requirements for exchange of active and reactive powers under different operating modes in the network with two voltage classes.

(2) Generally, the capacity of interconnecting transformer shall not be less than that of

the largest unit connected to the bus with two voltage classes, thus guaranteeing the conformance to the load requirements on the fault and maintenance sides through the interconnecting transformer in case of the faults or maintenance of the largest unit. In the meantime, the surplus capacity can be transmitted to another system as well through the interconnecting transformer in case of line maintenance or faults.

4) Substation Transformer

(1) The capacity of main transformer in the substation shall be selected based on the load planned for 5—10 years, with an appropriate consideration given to the long-term load development for 10—20 years. For suburban substations, the capacity of main transformer shall be combined with urban planning.

(2) The capacity of main transformer shall be determined based on the load nature of substation and the power grid structure.

(3) For the substations with important loads, it should be considered that the Class I and II loads of the user shall be guaranteed when one main transformer is shut down and the capacity of remaining transformers is within the allowed time after the capability of overload is calculated.

(4) For general substations, when one main transformer is shut down, the capacity of the remaining transformers shall be able to guarantee 70%—80% of the total load.

(5) For the substations in metropolitan suburbs, two main transformers are advisable in the substation if a looped network is formed at the medium and LV sides. For regional isolated substations or large industrial dedicated substations, the possibility of three main transformers shall be considered in the design. For the substations only installed with two main transformers as planned, the transformer foundation should be designed to the extent of being larger than the transformer capacity, so as to cope with the development of loads.

2. Transformer Form and Structure

1) Number of Phases

Three-phase and single-phase main transformers are mainly used, with the primary consideration of such factors as manufacturing conditions, reliability requirements and transportation conditions of transformer. In particular, for large transformers, the possibility of transportation shall be considered, to guarantee that the transportation clearance is not more than the allowed passage limits of tunnels, culverts and bridge openings, and that the transportation weight is not more than the allowed carrying capacities of pavements, bridges, vehicles and ships.

Compared with the three-phase transformer, the single-phase transformer is featured by large investment, large floor area, great operation losses, complex PDU structure, and heavy maintenance workload. However, the transportation of three single-phase transformers is more flexible with lower requirements than that of one three-phase transformer with the same capacity.

Therefore, the following principles shall be considered for the number of phases of the main transformer:

(1) The three-phase transformer shall be used at 330 kV and below if not subject to transport conditions.

(2) For the 500 kV transformers, three-phase transformers, two half-capacity three-phase transformers, or three single-phase transformers shall be selected after comparison of technology and economical efficiency.

2) Number of Windings

There are duplex-winding, three-winding or multi-winding power transformers based on the number of windings at each phase. The power transformers can be divided into ordinary duplex-winding, three-winding, auto- and LV winding splittypes based on electromagnetic structure.

Two duplex-winding or three-winding transformers can be adopted if the user is powered by two classes of boosted voltages in the power plant.

The three-winding transformer is mostly adopted for the power plants with the unit capacity of 125 MW and below. However, the through capacity of each winding of the three-winding transformer shall reach over 15% of the rated capacity. Otherwise, the winding has not been fully utilized, and adopting two duplex-winding transformers is more economical. In addition, at most 3 three-winding transformers are advisable in power plants or substations, avoiding the complex layout at the MV side, reducing the cost and decreasing the capacity of short circuit at the MV side.

For the power plants with the unit capacity over 200 MW, the "generator—duplex-winding transformer" unit is often connected to the system, and the interconnecting transformer is used between two classes of boosted voltages.

The three-winding interconnecting transformer is selected generally, and its LV winding can be connected to the HV station backup transformer or reactive power compensation device.

For the main transformer of expansion unit wiring, the LV split winding transformer is preferred, and the short-circuit current can be limited.

The auto-transformer is preferred if there are three voltage classes, i.e. 220 kV or above, which is featured by low loss, low price, high efficiency and low impedance.

3) Voltage Regulating Mode

The voltage must be maintained in the allowable range so as to guarantee the power quality of power grids. The number of turns in the HV winding of the transformer can be adjusted after adjusting the transformer tap switch, thus changing the ratio of transformation to adjust the voltage within a certain range. However, such method cannot cope with the voltage drop of power grid caused by insufficient reactive power capacity in the power grid.

There are two tap switching forms of transformer, i.e. no-load switching, known as no-load pressure regulation or exciterless pressure regulation, in the simple structure. However, the switching after power outage is only possible; and on-load switching, known as on-load pressure regulation, with the regulating range up to 30%, in the complicated structure and at a stiff price. However, voltage can be regulated without power outage in the following cases:

.(1) If connected to the main transformer of the power plant with significant changes in output power, especially when the direction of power flow is unfixed and the secondary voltage of transformer is required to be maintained at a certain level.

(2) If connected to the interconnecting transformer, which serves as the transmitting end or receiving end sometimes with reversible working characteristics, when the constant bus voltage is required so as to guarantee the power supply quality.

Generally, there are generators in the power plants, and the voltage can be regulated by excitation adjustment. Thus, on-load pressure regulation is less adopted for the main transformer.

4) Cooling Mode

The cooling modes of oil immersed transformer vary with different voltage classes, types and capacities. Generally, there are natural air cooling, forced air cooling, forced oil circulation air cooling, forced oil circulation water cooling, and forced oil circulation guided cooling.

For the medium-sized and small transformers with the capacity less than 31.5 MV·A, the flake or tube type radiation cooling on the transformer oil tank, the natural air cooling by electric fans, and the forced air cooling are adopted ususally.

For the transformers with the capacity of 31.5 MV·A or above, the forced oil circulation air cooling is adopted often, or the forced oil circulation water cooling can be adopted when the water is sufficient in the power plant.

For the super-huge transformers with the capacity of 350 MV·A or above, the forced oil circulation guided cooling is adopted in normal circumstances.

项目四　电气设备的选择

模块一　短路电流的效应

一、电气设备在运行中的两种工作状态

1. 正常工作状态
正常工作状态即电压和电流都不超过额定值的允许偏移范围，是一种长期工作状态。

2. 短路时工作状态
短路时工作状态即系统发生短路故障至故障切除的短时间内的工作状态。

短路电流大，持续时间短，认为在短路电流持续时间内所产生的全部热量都用来升高导体自身的温度，即认为是一个绝热过程。短路时导体温度变化范围很大，它的电阻和比热容不能再视为常数，而应为温度的函数。

二、短路电流的热效应

电气设备在工作过程中，由于自身存在着有功功率损耗，所以必然会引起电气设备的发热。

1. 电气设备运行中的三种损耗
（1）"铜损"，即载流导体的电阻损耗。
（2）"铁损"，即载流导体周围的金属构件（特别是铁磁物质）处于交变磁场中产生的磁滞和涡流损耗。
（3）"介损"，即绝缘材料在电场作用下产生的介质损耗。
这些损耗都转变成热量使电气设备的温度升高。

2. 电气设备工作时的两种发热及不良影响
根据导体通过电流的大小和持续时间长短的不同，可将导体发热分为长期发热和短路时发热两种。长期发热是指正常工作电流在较长时间内所引起的发热；短路时发热是指短路电流在极短的时间内所引起的发热。

发热不仅消耗能量，而且导致电气设备的温度升高，从而对电气设备产生不良的影响。如：
（1）使金属材料的机械强度下降。
（2）使导体接触部分的接触电阻增加。
（3）使绝缘材料绝缘性能下降。

3. 导体的发热和散热

（1）发热。

导体的发热主要来自导体电阻损耗的热量和太阳照射的热量。发热按流过电流的大小和时间分为长期发热和短时发热两大类。

① 长期发热指正常工作电流长期通过引起的发热。长期发热的热量，一部分散到周围介质中去，一部分使导体的温度升高。

② 短时发热是指短路电流通过时引起的发热。虽然短路的时间不长，但是短路电流很大，发热量很大，而且来不及散到周围介质中去，使导体温度迅速升高。

为了保证导体可靠的工作，规定了导体长期工作发热和短路时发热的温度限值，称为最高允许温度。裸导线长期工作时的最高允许温度一般为 70 ℃。在计及日照影响时，钢芯铝绞线及管型导体为 80 ℃；当导体接触面处有镀（搪）锡的可靠覆盖层时为 85 ℃；有银覆盖层时为 95 ℃。裸导体通过短路电流时的短时最高允许温度，对硬铝及铝锰合金为 200 ℃，对硬铜为 300 ℃。

（2）散热。

散热的过程实质是热量的传递过程，有三种形式：

① 导热。当物体的内部或相互接触的物体存在温差时，热量由高温区向低温区传递的过程。

② 对流。由温度不同的各部分流体发生相对运动将热量带走的过程。

③ 辐射。热量从高温物体以热射线方式传至低温物体的传播过程。

4. 大电流导体附近钢构的发热

大电流导体周围存在强大的交变磁场，使附近的钢构件中存在较大的磁滞和涡流损耗，钢构因而发热。

减小钢构损耗和发热的措施：

（1）加大钢构和载流导体之间的距离。

（2）断开载流导体附近的钢构闭合回路并加上绝缘垫。

（3）采用电磁屏蔽。

（4）采用分相封闭母线。

三、导体的电动力效应

所谓电动力是指载流导体在相邻载流导体产生的磁场中所受的电磁力。载流导体之间电动力的大小，取决于通过导体电流的数值、导体的几何尺寸、形状以及各相安装的相对位置等多种因素。

一般情况下，当电力系统中发生三相短路后，导体流过冲击短路电流时必然会在导体之间产生最大的电动力。

当电气设备通过短路电流时，短路电流所产生的巨大电动力对电气设备具有很大的危害性。如：

（1）载流部分可能因为电动力而振动，或者因电动力所产生的应力大于其材料允许应力而变形，其至使绝缘部件（如绝缘子）或载流部件损坏。

（2）电气设备的电磁绕组，受到巨大的电动力作用，可能使绕组变形或损坏。

（3）巨大的电动力可能使开关电器的触头瞬间解除接触压力，甚至发生斥开现象，导致设备故障。

1. 两根平行载流导体间的作用力

当两个平行导体通过电流时，由于磁场相互作用而产生电动力，电动力的方向与所通过的电流的方向有关。

当电流的方向相反时，导体间产生斥力；而当电流方向相同时，则产生吸力。

考虑导体截面形状和尺寸时，两平行导体间的电动力为

$$F = 2K_f i_1 i_2 \frac{L}{a} \times 10^{-7}$$

式中　i_1、i_2——通过导体的电流瞬时最大值，A；

　　　L——平行导体长度，m；

　　　a——导体轴线间距离，m；

　　　K_f——形状系数。

形状系数 K_f 与导体截面形状以及导体的相对位置有关。形状系数的确定较复杂。

2. 三相导体的电动力

三相短路时，每相导体所承受的电动力等于该相导体与其他两相之间电动力的矢量和。

三相导体布置在同一平面时，各相导体所通过的电流不同，故边缘相与中间相所承受的电动力也不同。最大冲击力发生在短路后 0.01 s，而且以中间相受力最大，电动力最大值为 $F_V^{(3)} = 1.73 i_k^{(3)} \frac{L}{a} \times 10^{-7}$。

两相短路时最大电动力小于同一地点三相短路时的最大电动力，故而要用三相短路时的最大电动力校验电气设备的动稳定。

如果导体的固有频率接近电动力中工频或二倍工频时，就会出现共振现象，甚至使导体及其构架损坏，凡连接发电机、主变压器以及配电装置中的导体需要考虑共振影响。

Program 4 Selection of Electrical Equipment

Module 1 Effect of Short-circuit Current

1. Two Working Conditions of Electrical Equipment during Operation

1) Normal Working Condition

Normal working condition is long-term service condition with the voltage and current within the allowable deviation range of rated values.

2) Working Condition During Short Circuit

Working condition during short circuit is the service condition of the system in a short period of time from the occurrence of short-circuit faults to the troubleshooting.

With heavy short-circuit current and short duration, it is regarded that the heat generated in the duration of short-circuit current is completely used to raise the temperature of the conductor itself, which is considered as an adiabatic process. In case of short circuit, the temperature of conductor varies widely. Its resistance and specific heat capacity shall be regarded as the function of temperature rather than constants.

2. Heat Effect of Short-circuit Current

The electrical equipment will heat inevitably due to its active power loss while working.

1) Three Losses of Electrical Equipment During Operation

(1) "Copper loss", the resistance loss of current carrying conductor.

(2) "Iron loss", the hysteresis and eddy current losses of the metal components (especially ferromagnetic materials) around the current carrying conductor in the AC magnetic field.

(3) "Dielectric loss", the dielectric loss generated by insulating materials under the electric field action.

These losses are converted into heat, causing the temperature rise of electrical equipment.

2) Two Types of Heating and Adverse Effects of Electrical Equipment While Working

The heating of conductor can be divided into long-term heating and short-circuit heating based on the size and duration of the current through the conductor. Long-term heating refers to the heating caused by normal working current for a long period of time. Short-circuit heating refers to the heating caused by short-circuit current in an extremely short period of time.

Heating will consume energy, and also cause the temperature rise of electrical equipment, resulting in adverse effects on electrical equipment. E.g.:

(1) Reduced mechanical strength of metal materials.

(2) Increased contact resistance of conductor contact segments.

(3) Lowered insulating property of insulating materials.

3) Heat and Heat Dissipation of Conductors

(1) Heating.

The conductor mainly heats up due to the heat from the resistance loss of conductor and the exposure to sunshine. Heating is divided into long-term heating and short-term heating based on the let-through current size and duration.

① Long-term heating refers to the heating caused by the long-term passing of normal working current. The heat generated for a long period of time partially disperses into the surrounding media and partially increases the temperature of conductor.

② Short-term heating refers to the heating caused by the passing of short-circuit current. Although the short circuit lasts for a short time, the short-circuit current is very high and the heat generated is great, without enough time to dissipate the heat into the surrounding media, resulting in fast conductor temperature rise.

In order to guarantee the reliable conductor operation, the temperature limits, known as the maximum allowable temperature, have been specified for the long-term heating of conductor under operating conditions and the heating at short circuit. The maximum allowable temperature is generally 70 °C for bare wires while working for a long period of time. The maximum allowable temperature is 80 °C for aluminium cable steel reinforced and tubular conductor when the sunshine exposure effect is considered. The maximum allowable temperature is 85 °C and 95 °C respectively when there are tin-plated (tinned) reliable coating and silver coating at the contact surface of conductor. The maximum allowable short-term temperature is 200 °C and 300 °C for duralumin/aluminum-manganese alloy and for hard copper respectively when the bare conductor passes through the short-circuit current.

(2) Heat dissipation.

The process of heat dissipation is equivalent to the transfer of heat in essence, consisting of three forms:

① Heat conduction. The process in which the heat is transferred from high temperature zone to low temperature zone in case that the temperature difference exists in the object or the objects in contact with each other.

② Convection. The process in which the heat is carried away through the relative motion of fluids at different temperatures.

③ Radiation. The process in which the heat is transferred from the high-temperature object to the low-temperature object through heat rays.

4) Heating of Steel Structures Near High Current Conductors

A strong AC magnetic field exists around the high current conductor, leading to significant hysteresis and eddy current losses in the nearby steel components. The steel structure heats up on this account.

Measures to reduce steel structure loss and heat are as follows:

(1) Increase the distance between steel structure and current carrying conductor.

(2) Disconnect the closed circuit of steel structure near the current carrying conductor, and add insulation pads.

(3) Adopt the electromagnetic shielding.

(4) Utilize the split-phase enclosed buses.

3. Electrodynamic Force of Conductor

The mentioned electrodynamic force refers to the electromagnetic force borne by the current carrying conductor in the magnetic field generated by adjacent current carrying conductors. The electrodynamic force between current carrying conductors depends on multiple factors, such as value of current passing through conductor, geometric size and shape of conductor, and relative installation position of each phase.

Generally, in case of the three-phase short circuit in the power system, the maximum electrodynamic force will be generated inevitably between conductors when the conductor passes through the impulse short-circuit current.

When the electrical equipment passes through the short-circuit current, the huge electrodynamic force generated by short-circuit current is greatly harmful to electrical equipment. E.g.:

(1) The carrying current may vibrate due to electrodynamic force, or deform due to the stress from electrodynamic force exceeding the allowable stress of its materials, or even damage the insulating components (such as insulators) or current carrying components.

(2) The electromagnetic winding of electrical equipment may make the winding deformed or damaged due to the huge electrodynamic force suffered.

(3) The huge electrodynamic force may lead to the instantaneous release of contact pressure for the contact terminal of switching device, and even cause the repulsion, resulting in equipment faults.

1) Acting Force between Two Parallel current Carrying Conductors

When the current passes through two parallel conductors, the direction of electrodynamic force from magnetic field interaction is related to that of the current that passes through.

Repulsion occurs between conductors when the current directions are opposite. Suction occurs when the current directions are the same.

In consideration of the cross-section shape and size of the conductor, the electrodynamic force between two parallel conductors is as follows:

$$F = 2K_f i_1 i_2 \frac{L}{a} \times 10^{-7}$$

Where, i_1 and i_2—maximum instantaneous value of current passing through the conductor, A;

L—length of parallel conductor (m);

a—Distance between conductor axes (m);

K_f—form factor.

The form factor K_f is related to the cross-section shape and relative position of the conductor. It is relatively complex to determine the form factor.

2) Electrodynamic Force of Three-phase Conductor

In case of three-phase short circuit, the electrodynamic force borne by the conductor of each phase is equal to the vector sum of the electrodynamic force between the conductor of this phase and the conductors of other two phases.

When the three-phase conductors are arranged in the same plane, the currents passing through the conductors of all phases are different. Therefore, the electrodynamic forces borne at the edge phase and the intermediate phase are also different. The maximum impact force occurs within 0.01s after short circuit, and the force acting on the intermediate phase is the highest. The maximum value of electrodynamic force is $F_V^{(3)} = 1.73 i_k^{(3)} \frac{L}{a} \times 10^{-7}$.

The maximum electrodynamic force during the two-phase short circuit is less than that during three-phase short circuit at the same location. Thus, dynamic stability of electrical equipment shall be verified by the maximum electrodynamic force during three-phase short circuit.

If the inherent frequency of conductor is approximate to the power frequency or doubled power frequency of electrodynamic force, resonance may occur, and even the conductor and its frame will be damaged. Resonance effect shall be considered for the conductors connected to generators, main transformers and PDUs.

模块二　电气设备选择的一般条件

为了保障高压电气设备的可靠运行，必须按正常工作条件进行选择，并按短路状态进行校验。校验的短路电流一般取三相短路时短路电流，若发电机出口的两相短路，或中性点直接接地系统及自耦变压器等回路中的单相、两相接地短路较三相短路严重时，则应按严重情况校验。

高压电气设备选择与校验的一般条件有：

（1）按正常工作条件包括电压、电流、频率、开断电流等选择。

（2）按短路条件包括动稳定、热稳定校验。

（3）按环境工作条件如温度、湿度、海拔等选择。

一、按正常工作条件选择

1. 按照额定电压选择

电气设备的最高工作电压允许值不得低于所接电网的最高运行电压。常规定一般电气设备允许的最高工作电压为其额定电压的 1.1~1.15 倍,而电网运行电压的波动幅值一般不超过 1.15 倍的电网额定电压。因此,一般按照电气设备额定电压 U_N 不低于设备所接电网额定电压 U_{Ns} 的条件选择电气设备,即

$$U_N \geqslant U_{Ns}$$

2. 按照额定电流选择

电气设备的额定电流为额定环境条件(温度、海拔高度、风速、日照、光照条件等)下,电气设备的长期允许通过电流。运行中的电气设备额定电流 I_N 一般不小于所在回路在各种合理运行方式下的最大持续工作电流 I_{max},即

$$I_N \geqslant I_{max}$$

二、按短路条件校验

用熔断器保护的电气设备可不验算热稳定。当熔断器有限流作用时,可不验算动稳定。用熔断器保护的电压互感器回路,可不验算动、热稳定。不流通电流的支柱绝缘子不校验热稳定。电缆具有足够强度,可不校验动稳定。

1. 短路热稳定校验

(1)导体。

若按正常工作条件选择的导体截面积为 S,则满足热稳定的条件为

$$S > S_{min}$$

式中,S_{min} 为按热稳定确定的导体最小截面积,单位为 mm^2。

(2)电气设备。

短路电流通过电气设备时,要求电气设备各部件温度(或发热效应)应不超过允许值。若电气设备允许通过的热稳定电流 I_t 和允许持续时间 t,则满足热稳定的条件为

$$I_t^2 t \geqslant Q_k$$

式中,Q_k 为短路电流产生的热效应,单位为 $(kA)^2 \cdot s$。

2. 短路动稳定校验

(1)硬导体。

若硬导体材料的最大允许应力为 σ_{a_1},则满足动稳定的条件为

$$\sigma_{a_1} \geqslant \sigma_{max}$$

式中,σ_{max} 为导体的最大计算应力,单位为 Pa。

(2)电气设备。

动稳定是电气设备承受短路电流机械效应的能力。若电气设备允许通过的动稳定电

流及其有效值 i_{es}，则满足动稳定的条件为

$$i_{es} \geq i_{sh} \text{ 或 } I_{es} \geq I_{sh}$$

式中，i_{sh} 和 I_{sh} 分别为实际计算得到的短路电流最大冲击值及其有效值，单位为 kA。

表 4-1 高压电气设备选择与校验项目

设备名称	额定电压	额定电流	额定开断电流	短路校验		环境条件
				动稳定	热稳定	
断路器	√	√	√	√	√	√
隔离开关	√	√		√	√	√
负荷开关	√	√		√	√	√
熔断器	√	√	√			
电流互感器	√	√		√	√	√
电压互感器	√					√
绝缘子	√			√		√
穿墙套管	√	√		√	√	√
母线				√	√	√
电缆	√				√	√

三、按照环境工作条件选择

（1）按照安装地点、使用条件和运检要求选择电气设备的种类和型式。

（2）考虑实际环境温度、海拔高度、风速、光照、污染情况等环境条件的影响。当超过一般电气设备的使用条件时（如高原地带、强风地区、污染严重地区等），应选择适用此环境条件的电气设备，并采取相应措施。

Module 2　General Conditions for Selection of Electrical Equipment

In order to ensure the reliable operation of HV electrical equipment, electrical equipment must be selected according to normal working conditions and verified at the short-circuit status. The short-circuit current in case of three-phase short circuit is selected generally for validation. Validation shall be subject to the seriousness in case of two-phase short circuit at the generator outlet, or if the single-phase and two-phase grounded short circuits in the circuits of neutral point directly grounded system and the auto-transformer are more serious than the three-phase short circuit

General conditions for the selection and verification of HV electrical equipment are as follows:

(1) Selection subject to normal working conditions, including voltage, current, frequency and breaking current.

(2) Verification subject to short-circuit conditions, including dynamic stability and thermal stability.

(3) Selection subject to environmental working conditions, such as temperature, humidity and altitude.

1. Selection Subject to Normal Working Conditions

1) Selection Subject to Rated Voltage

The maximum allowable working voltage of electrical equipment cannot be lower than the maximum operating voltage of the power grid connected. Generally, it is specified that the maximum allowable working voltage of general electrical equipment is 1.1—1.15 times its rated voltage, while the fluctuation amplitude of the operating voltage of power grid is not more than 1.15 times the rated voltage of power grid generally. Therefore, the electrical equipment shall be selected under the condition that the rated voltage U_N of electrical equipment is not lower than the rated voltage U_{NS} of the power grid connected by the equipment generally.

$$U_N \geqslant U_{NS}$$

2) Selection Subject to Rated Current

The rated current of electrical equipment refers to the long-term allowable let-through current of electrical equipment under rated environmental conditions (in terms of temperature, altitude, wind speed, sunlight and lighting). The rated current I_N of electrical equipment in service is generally not less than the maximum continuous working current I_{max} of the circuit under all reasonable operating modes, i.e:

$$I_N \geqslant I_{max}$$

2. Validation Subject to Short-circuit Conditions

Thermal stability may not be calculated for the electrical equipment protected by fuses. Dynamic stability may not be calculated in case of current limiting by the fuse. Checking calculation is not required for thermal stability and dynamic stability for the circuits of voltage transformer protected by the fuse. Thermal stability may not be calculated for the post insulator without current through. Dynamic stability may not be calculated for the cables with sufficient strength.

1) Short-circuit Thermal Stability Validation

(1) Conductor.

If the cross-sectional area of the conductor selected under normal working conditions is S, the conditions under which thermal stability requirements are met are as follows:

$$S > S_{min}$$

Where, S_{min}—minimum cross-sectional area of the conductor determined subject to thermal stability, mm².

(2) Electrical equipment.

When the short-circuit current passes through electrical equipment, the temperature (or heating effect) of each component of the electrical equipment must not exceed the allowable value. For the thermal stability current I_t allowed to pass through the electrical equipment and the allowable duration t, the conditions under which thermal stability requirements are met are as follows:

$$I_t^2 t \geqslant Q_k$$

Where, Q_k—heat effect from short-circuit current, (kA)² · s.

2) Short-circuit Dynamic Stability Validation

(1) Rigid conductor.

For the maximum allowable stress σ_{al} of the rigid conductor material, the conditions under which dynamic stability requirements are met are as follows:

$$\sigma_{al} \geqslant \sigma_{max}$$

Where, σ_{max}—maximum calculated stress of conductor, Pa.

(2) Electrical equipment.

Dynamic stability refers to the ability of electrical equipment to withstand the mechanical effects of short-circuit currents. For the dynamic stability current i_{es} allowed to pass through the electrical equipment and its effective value i_{es}, the conditions under which dynamic stability requirements are met are as follows:

$$i_{es} \geqslant i_{sh} \text{ or } I_{es} \geqslant I_{sh}$$

Where, i_{sh} and I_{sh}—the actually calculated maximum impact value of short-circuit current and its effective value respectively, kA.

Table 4-1 HV Electrical Equipment Selection and Calibration Items

Equipment name	Rated voltage	Rated current	Rated breaking current	Short circuit validation		Environmental conditions
				Dynamic stability	Thermal stability	
Circuit breaker	√	√	√	√	√	√
Disconnector	√	√		√	√	√
Load switch	√	√		√	√	√
Fuse	√	√	√			√
Current transformer	√	√		√	√	√
Voltage transformer	√					√
Insulator	√			√		√
Wall-through bushing	√	√		√	√	√
Bus				√	√	√
Cable	√				√	√

3. Selection Subject to Environmental Working Conditions

(1) The type and model of electrical equipment shall be selected as required by installation site, service conditions and operating inspection.

(2) The impact of environmental conditions, including actual ambient temperature, altitude, wind speed, lighting and pollution, shall be considered. Beyond the service conditions of general electrical equipment (such as plateaus, areas with strong wind, and heavily polluted areas), the electrical equipment suitable for such environmental conditions shall be selected and corresponding measures shall be taken.

模块三　高压电气设备的选择

一、高压开关电器的选择

选择高压断路器、高压隔离开关和高压负荷开关的长期工作条件基本相同，区别在于它们短路校验的内容不同，如隔离开关和负荷开关不校验短路开断电流。

1. 种类和型式的选择

根据用途、安装地点、安装方式、结构类型和价格因素等综合条件进行合理选择。

2. 额定电压选择

开关电器的额定电压应等于或大于安装地点电网的额定电压，即

$$U_N \geqslant U_{Ns}$$

3. 额定电流选择

开关电器的额定电流应等于或大于通过断路器的长期最大负荷电流，即

$$I_N \geqslant I_{max}$$

4. 断路器的开断电流选择

断路器的允许开断电流 I_{Nbr} 应大于或等于断路器实际开断时间的三相短路电流周期分量有效值 I_{ap}，即

$$I_{Nbr} \geqslant I_{ap}$$

5. 动稳定校验

开关电器允许的动稳定电流峰值 i_{es} 应大于或等于流过断路器的三相短路冲击电流 i_k，即

$$i_{es} \geqslant i_k$$

6. 热稳定校验

开关电器 t 秒钟热稳定电流 I_t 算出的允许热效应大于或等于通过断路器的短路电流热效应,即

$$I^2t \geqslant Q_k$$

二、高压熔断器的选择

高压熔断器按额定电压、额定电流、开断电流和选择性等项来选择和校验。

1. 额定电压选择

对于一般的高压熔断器,其额定电压 U_N 必须大于等于电网的额定电压 U_{Ns},即

$$U_N \geqslant U_{Ns}$$

2. 额定电流选择

熔断器的额定电流选择,包括熔断器熔管的额定电流和熔体的额定电流的选择。

(1) 熔管额定电流的选择。

为了保证熔断器壳不致损坏,高压熔断器的熔管额定电流 I_{Nft} 应大于或等于熔体的额定电流 I_{Nf},即

$$I_{Nft} \geqslant I_{Nf}$$

(2) 熔体额定电流选择。

为了防止熔体在通过变压器励磁涌流和保护范围以外短路及电动机自启动等冲击电流时误动作,保护 35 kV 及以下电力变压器的高压熔断器,其熔体的额定电流选择式为

$$I_{Nf} = KI_{max}$$

式中 I_{max}——电力变压器回路最大工作电流;

K——可靠系数(不计电动机自启动时 $K=1.1\sim1.3$,考虑电动机自启动时 $K=1.5\sim2.0$)。

用于保护电力电容器的高压熔断器的熔体,当系统电压升高或波形畸变引起回路电流增大或运行过程中产生涌流时不应误熔断,其熔体选择式为

$$I_{Nf} = KI_{Nc}$$

式中:I_{Nc}——电力电容器回路的额定电流;

K——可靠系数(对限流式高压熔断器,当一台电力电容器时 $K=1.5\sim2.0$,一组电力电容器时 $K=1.3\sim1.8$)。

3. 熔断器开断电流校验

$$I_{Nbr} \geqslant I_k \text{(或 } I'' \text{)}$$

对于没有限流作用的熔断器,选择时用短路电流的有效值 I_k 进行校验;对于有限流作用的熔断器,在电流达最大值之前已截断,故可不计非周期分量影响,而采用 I'' 进行校验。

4. 熔断器选择性校验

为了保证前后两级熔断器之间或熔断器与电源（或负荷）保护装置之间动作的选择性，应进行熔体选择性校验。各种型号熔断器的熔体熔断时间可由制造厂提供的安秒特性曲线上查出。

三、互感器的选择

（一）电流互感器的选择

1. 额定电压和电流选择

电流互感器的一次额定电压和电流必须满足：

$$U_N \geqslant U_{Ns}$$

$$I_N \geqslant I_{max}$$

式中　U_{Ns}——电流互感器所在电力网的额定电压，kV；
　　　U_N、I_N——电流互感器的一次额定电压和电流；
　　　I_{max}——电流互感器一次回路最大工作电流，A。

2. 电流互感器种类和型式选择

在选择互感器时，应根据安装地点（如屋内、屋外）和安装方式（如穿墙式、支持式、装入式等）选择型式。

3. 选择电流感器的准确度等级和额定容量

为了保证仪表的准确度，互感器的准确度等级不得低于所供测量仪表的准确度等级。当所供仪表要求不同准确度等级时，应按最高级别来确定互感器的准确级。

为了保证互感器的准确度等级，互感器二次侧所接的最大负荷 S_2 应不大于该准确度等级所规定的额定容量 S_{N2}，即

$$S_{N2} \geqslant S_2 = I_{N2}^2 Z_{2L}$$

4. 热稳定校验

电流互感器热稳定能力常以 1s 允许通过一次额定电流 I_{N1} 的倍数 K_t 来表示，故热稳定校验式为

$$(K_t I_{N1})^2 \geqslant I_k^2 t_k \text{（或} \geqslant Q_k\text{）}$$

式中　I_k——短路电流稳态值；
　　　t_k——短路计算时间。

5. 动稳定校验

电流互感器常以允许通过一次额定电流最大值（$\sqrt{2} I_{N1}$）的倍数 K_{es} 稳定电流倍数，表示其内部动稳定能力，所以内部动稳定校验式为

$$\sqrt{2}I_{N1}K_{es} \geq i_k$$

(二) 电压互感器的选择

1. 按一次回路电压选择

为了确保电压互感器安全和在规定的准确度等级下运行，电压互感器一次绕组所接电力网电压 U_{Ns} 应在 $0.85 \sim 1.16 U_{N1}$ 范围内变动，即满足下列条件：

$$0.85 U_{N1} < U_{Ns} < 1.16 U_{N1}$$

2. 按二次回路电压的选择

二次回路电压必须满足保护和测量使用标准仪表的要求，根据电压互感器接线的不同，二次电压各不相同，选择时可按表 4-2 选择。

表 4-2 电压互感器额定电压选择

形式	一次电压/V		二次电压/V	第三绕组电压/V	
单相	接于一次线电压上（如 Vv 接法）	U_{Ns}	100		
	接于一次相电压上	$U_{Ns}\sqrt{3}$	$100/\sqrt{3}$	中性点非直接接地系统	$100/3$、$100/\sqrt{3}$
				中性点直接接地系统	100
三相	U_{Ns}		100	$100/3$	

注：U_{Ns} 为系统额定电压。

3. 种类和型式选择

电压互感器的种类和型式应根据安装地点和使用条件进行选择。例如：在 6～35 kV 屋内配电装置中一般采用油浸式或浇注式；110～220 kV 配电装置，一般采用串级式电磁式电压互感器；在 200 kV 及其以上配电装置，当容量和准确度等级满足要求时，一般采用电容式电压互感器。

4. 按容量和准确度等级选择

有关电压互感器准确度等级选择应满足所供测量仪表的最高准确度等级，应根据仪表和继电器接线要求选择电压互感器的接线方式，并尽可能将负荷均分布在各相上，然后计算各相负荷大小。

互感器的额定二次容量（对应于所要求的准确度等级）S_{N2}，应不小于互感器的二次负S_2，即 $S_{N2} \geq S_2$，且

$$S_2 = \sqrt{\left(\sum S_{me}\cos\varphi\right)^2 + \left(\sum S_{me}\sin\varphi\right)^2} = \sqrt{\left(\sum P_{me}\right)^2 + \left(\sum Q_{me}\right)^2}$$

式中 S_{me}、P_{me}、Q_{me}——各仪表的视在功率、有功功率、无功功率；
$\cos\varphi$——各仪表的功率因数。

四、母线、电缆、绝缘子的选择

(一) 母线的选择

1. 母线材料、类型和布置方式

硬母线的材料有铜、铝、钢，一般采用铝或铝合金作为母线材料。常用的软导线有钢芯铝绞线、组合导线、分裂导线和扩径导线，多用于户外配电装置。

母线的截面形状有矩形、槽形和管形。矩形导体一般只用于 35 kV 及以下，电流在 4 000 A 及以下的配电装置中。槽形导体一般用于 4 000~8 000 A 的配电装置中。管形导体用于 8 000 A 以上的大电流母线，或用在 110 kV 及以上的配电装置中。

母线的布置方式应根据载流量的大小、短路电流水平和配电装置的具体情况而定。

2. 母线截面选择

母线截面可按长期发热允许电流或经济电流密度选择。除配电装置的汇流母线外，对于年负荷利用小时数大、传输容量大、长度在 20 m 以上的导体，其截面一般按经济电流密度来选择。

(1) 按导体长期发热允许电流选择。其计算式为

$$KI_{a1} \geqslant I_{max}$$

式中　I_{max}——导体所在回路中的最大持续工作电流；

　　　I_{a1}——在额定环境温度 θ_0=25℃时导体允许电流；

　　　K——与实际温度和海拔有关的综合修正系数。

(2) 按经济电流密度选择。

按经济电流密度选择导体截面可使年计算费用最低。对不同的导体种类和不同的最大负荷利用小时数 T_{max}，将有一个年计算费用最低的电流密度，称为经济电流密度 J。

导体的经济截面：

$$S = \frac{I_{max}}{f}$$

式中　I_{max}——正常工作时的最大持续工作电流。

3. 电晕电压校验

电晕放电将引起电晕损耗、无线电干扰、噪声和金属腐蚀等许多不利影响。对于 110 kV 及以上裸导体可按晴天不发生全面电晕条件校验，即裸导的临界电压 U_{cr} 应大于最高工作电压 U_{max}，即

$$U_{cr} > U_{max}$$

当所选软导线型号和管形导体外径大于或等于下列数值时，可不进行电晕校验：110 kV，LGJ-70/ϕ20；220kY，LGJ-300/ϕ30。

4. 热稳定校验

在校验导体热稳定时，若计及集肤效应系数 K_s 的影响，由热稳定决定的导体最小截面为

$$S_{\min} = \sqrt{Q_k K_s /(A_k - A_i)} = \sqrt{Q_k K_s / C}$$

式中，C 为热稳定系数，$C = A_k - A_i$，C 值与导体材料及工作温度有关，所选截面应大于等于 $S_{\min}$。

5. 硬导体的动稳定

各种形状的硬导体通常都安装在支柱绝缘子上，短路冲击电流产生的电动力将使导体发生弯曲甚至造成损坏。因此，硬导体应按弯曲情况进行应力计算，而软导体不必进行动稳定校验。

若每相不少于两条导体，当短路电流通过导体时，导体的横截面受到相间弯矩 M_{ph} 的作用产生相间应力 σ_{ph}，同时受到条间弯矩 M_b 的作用产生条间应力 σ_b。

当 σ_{ph} 和 σ_b 方向相同时，产生的最大应力 $\sigma_{\max}$ 为

$$\sigma_{\max} = \frac{M_{ph}}{W_{ph}} + \frac{M_b}{W_b} = \sigma_{ph} + \sigma_b$$

式中，W_{ph} 和 W_b 分别为导体相间和条间抗弯曲截面系数。

当 $\sigma_{\max}$ 不超过导体最大允许应力 σ_{al} 时，则导体满足动稳定，即

$$\sigma_{\max} \leqslant \sigma_{al}$$

6. 导体共振校验

对于重要回路（如发电机、变压器及汇流母线等）的导体应进行共振校验。按下式计算：

$$f_1 = \frac{N_f}{L^2}\sqrt{\frac{EI}{m}}$$

式中　f_1——一阶固有频率，Hz；

L——跨距，m；

N_f——频率系数，N_f 根据导体连续跨数和支撑方式而异；

E——导体的弹性模量，P_a；

I——导体断面二次距，m^4。

（1）已知绝缘子跨距。

当硬导体的一阶固有频率 f_1 位于共振频率范围内时，查出 β 值；当 f_1 超出共振频率范围时，$\beta \approx 1$。

（2）未知绝缘子跨距。

当所选绝缘子实际跨距 L 不超过硬导体不发生共振时允许的最大绝缘子跨距 $L_{\max}$ 时，$\beta \approx 1$。

（二）电缆的选择

电力电缆应按下列条件选择和校验：① 电缆芯线材料及型号；② 额定电压；③ 截面选择；④ 允许电压降校验；⑤ 热稳定校验。电缆的动稳定由厂家保证，可不必校验。

1. 电缆芯线材料及型号选择

电力电缆芯线有铜芯和铝芯两种，国内工程一般选用铝芯电缆。电缆的型号很多，应根据其用途、敷设方式和使用条件进行选择。除 110 kV 及以上采用单相交联聚乙烯电缆或单相高压充油电缆外，一般采用三相铝芯油浸纸绝缘电缆、橡皮绝缘电缆、聚氯乙烯绝缘电缆或交联聚乙烯电缆；动力电缆通常采用三芯或四芯（三相四线）；高温场所宜用耐热电缆；重要直流回路或保安电源电缆宜选用阻燃型电缆；直埋地下一般选用钢带铠装电缆；潮湿或腐蚀地区应选用塑料护套电缆；敷设在高落差大的地点，应采用交联聚乙烯电缆。随着材料技术的发展，阻燃耐热型交联聚乙烯电缆得到了越来越广泛的应用。油浸纸绝缘和充油电缆等已趋于淘汰。

2. 电压选择

电缆的额定电压 U_N 应大于等于所在电网的额定电压 U_{Ns}，即

$$U_N \geqslant U_{Ns}$$

3. 截面选择

电力电缆截面一般按长期发热允许电流选择，当电缆的最大负荷利用小时 T_{max}>5 000 h，且长度超过 20 m 时，则应按经济电流密度选择。电缆截面选择方法与裸导体基本相同。

为了不损伤电缆绝缘及保护层，敷设时电缆应保持一定的弯曲半径，如多芯纸绝缘铅包电缆的弯曲半径不应小于电缆外径的 15 倍。

4. 允许电压降校验

对供电距离较远、容量较大的电缆线路，应校验其电压损失 $\Delta U\%$。一般应满足 $\Delta U\%<5\%$。对于三相交流，其计算公式为

$$\Delta U\% = 173 I_{max} L (r\cos\varphi + x\sin\varphi)/U$$

式中　U、L——线路工作电压（线电压）和长度；
　　　$\cos\varphi$——功率因数；
　　　r、x——单位长度的电阻和电抗。

5. 热稳定校验

由于电缆芯线一般系多股线构成，截面在 400 mm² 以下时，$K\approx 1$，满足电缆热稳定的最小截面可以简化写成：

$$S_{\min} = Q_k/C$$

（三）绝缘子的选择

支柱绝缘子应按额定电压和类型选择，并进行短路时动稳定校验。穿墙套管应按额定电压、额定电流和类型选择，按短路条件校验动、热稳定。

1. 按额定电压选择支柱绝缘子和穿墙套管

支柱绝缘子和穿墙套管的额定电压 U_N 应大于等于所在电网的额定电压 U_{Ns}，即

$$U_N > U_{Ns}$$

2. 按额定电流选择穿墙套管

穿墙套管的额定电流 I_N。应大于等于回路中最大持续工作电流，即

$$I_N > KI_{\max}$$

式中　K——温度修正系数。

对母线型穿墙套管，因本身无导体，不必按此项选择和校验热稳定，只需保证套管的型式穿过母线的尺寸相配合。

3. 支柱绝缘子和套管的种类和型式选择

根据装置地点、环境选择屋内、屋外（或防污式）及满足使用要求的产品型式。

4. 穿墙套管的热稳定校验

套管耐受短路电流的热效应 $I_t^2 t$ 应大于或等于短路电流通过套管所产生的热效应 Q_k，即

$$I_t^2 t \geqslant Q_k$$

5. 支柱绝缘子和套管的动稳定校验

支柱绝缘子和穿墙套管均需进行动稳定校验。若短路时，支柱绝缘子（或穿墙套管）的受力为其相邻跨导体上电动力 $F_{\max}$ 的平均值，则支柱绝缘子（或穿墙套管）的抗弯破坏强度 F_{de} 应满足：

$$0.6 F_{de} \geqslant \frac{H_1}{H} F_{\max}$$

Module 3　Selection of HV Electrical Equipment

1. Selection of HV Switching Device

The long-term working conditions are basically the same for selecting HV circuit breakers, HV disconnectors and HV load switches, while the difference lies in diversified contents of short circuit validation. For example, the short-circuit breaking current will not be verified for disconnectors and load switches.

1) Selection of Type and Model

Reasonable selection shall be subject to comprehensive conditions, such as purposes, installation site, installation methods, structural type and price factor.

2) Rated Voltage Selection

The rated voltage of switching device shall be equal to or more than the rated voltage of the power grid at the installation site, i.e.:

$$U_N \geqslant U_{Ns}$$

3) Rated Current Selection

The rated current of switching device shall be equal to or more than the long-term maximum load current through the circuit breaker, i.e.:

$$I_N \geqslant I_{max}$$

4) Breaking Current Selection of Circuit Breaker

The allowable breaking current I_{Nbr} of the circuit breaker shall be more than or equal to the effective value I_{ap} of the three-phase short-circuit current periodic component of the actual breaking time of circuit breaker, i.e.:

$$I_{Nbr} \geqslant I_{ap}$$

5) Dynamic Stability Validation

The allowable peak dynamic stable current i_{es} of the switching device shall be more than or equal to the three-phase short-circuit impulse current i_k through the circuit breaker, i.e.:

$$i_{es} \geqslant i_k$$

6) Thermal Stability Validation

The allowable heat effect $I_t^2 t$ calculated by the thermal stability current I_t of the switching device at t second(s) shall be more than or equal to the heat effect of the short-circuit current through the circuit breaker, i.e.:

$$I_t^2 t > Q_k$$

2. Selection of HV Fuses

HV fuses are selected and calibrated based on rated voltage, rated current, breaking current and selectivity.

1) Rated Voltage Selection

For general HV fuses, the rated voltage U_N must be more than or equal to the rated voltage U_{Ns} of the power grid, i.e.:

$$U_N \geqslant U_{Ns}$$

2) Rated Current Selection

For the rated current selection of the fuse, the rated currents of fuse cartridge and melt are selected.

(1) Rated current selection of cartridge.

In order to guarantee no damage to the fuse housing, the rated current I_{Nft} of the cartridge of the HV fuse shall be more than or equal to the rated current I_{Nf} of the melt, i.e.:

$$I_{Nft} \geqslant I_{Nf}$$

(2) Rated current selection of melt.

In order to prevent the malfunction of the melt when the excitation surge current of transformer passes through and the short-circuit and motor self-starting impulse currents beyond the protective range pass through, and in order to protect the HV fuses of the 35 kV and below power transformers, the selection formula for the rated current of melt is as follows:

$$I_{Nf} = KI_{max}$$

Where, I_{max}—maximum working current of power transformer circuit;

K—reliability coefficient (K=1.1—1.3 if the self-starting of motor is not considered, while K=1.5—2.0 if that is considered).

For the melt of the HV fuse used to protect the power capacitor, fusing by mistake is not allowed when the system voltage rises, or the waveform distortion increases the loop current, or the inrush current occurs during operation. The melt selection formula is as follows:

$$I_{Nf} = KI_{Nc}$$

Where, I_{Nc}—rated current of power capacitor circuit;

K—reliability coefficient (for the current-limiting HV fuse, K=1.5—2.0 in case of one power capacitor, and K=1.3—1.8 in case of one set of power capacitors).

3) Breaking Current Validation of Fuse

$$I_{Nbr} \geqslant I_k \text{ (or } I'')$$

For the fuses without current limiting, the effective value I_k of the short-circuit current shall be selected for validation. For the fuses with current limiting, the current has been cut off before reaching the maximum value. Thus, the influence of the non-periodic component is not considered, but I'' is adopted for validation.

4) Selectivity Validation of Fuse

Selectivity validation is required for the melt in order to ensure the selectivity of action between the front-stage and backward-stage fuses or between the fuse and the power supply (or load) protection device. The fusing time of the melts on various types of fuses can be checked on the ampere-second characteristic curve provided by the manufacturer.

3. Selection of Transformers

1) Selection of Current Transformer

(1) Selection of rated voltage and current.

The primary rated voltage and current of the current transformer must meet the following requirements:

$$U_N \geqslant U_{Ns}$$

$$I_N \geqslant I_{max}$$

Where, U_{Ns}—rated voltage of the power grid in which the current transformer is located (kV);

U_N and I_N—primary rated voltage and current of current transformer;

I_{max}—maximum working current of the primary circuit of current transformer (A).

(2) Selection of current transformer type and model.

The type of transformer shall be selected based on the installation site (such as indoor or outdoor) and the installation method (such as wall-through, supporting and built-in).

(3) Selection of accuracy class and rated capacity of current transformer.

In order to guarantee the instrument accuracy, the accuracy class of the transformer cannot be lower than that of the provided measuring instrument. When different accuracy classes are required for the provided instruments, the accuracy class of the transformer shall be determined subject to the highest class.

In order to guarantee the accuracy class of transformer, the maximum load S_2 connected at the secondary side of the transformer shall not exceed the rated capacity S_{N2} specified by the accuracy class, i.e.:

$$S_{N2} \geqslant S_2 = I_{N2}^2 Z_{2L}$$

(4) Thermal Stability Validation.

The thermal stability capability of current transformer is often indicated by the multiple K_t of the primary rated current I_{N1} within 1s.

Therefore, the thermal stability validation formula is as follows:

$$(K_t I_{N1})^2 \geqslant I_K^2 t_K \,(\text{or} \geqslant Q_K)$$

Where, I_k—steady-state value of short-circuit current;

t_k—short-circuit calculation time.

(5) Dynamic stability validation.

The internal dynamic stability capability of the current transformer is often indicated by the multiple K_{es} (dynamic stability current multiple) of the maximum primary rated current ($\sqrt{2}I_{N1}$) allowed to pass through. Therefore, the internal dynamic stability validation formula is as follows:

$$\sqrt{2}I_{N1}K_{es} \geq i_k$$

2) Selection of Voltage Transformer

(1) Selection subject to primary circuit voltage.

In order to guarantee the safety of voltage transformer and its operation under the specified accuracy class, the voltage U_{Ns} of the power grid connected to the primary winding of the voltage transformer shall vary within the range of 1.16—0.85U_{NI}, i.e. meeting the following conditions:

$$0.85U_{NI} < U_{Ns} < 1.16U_{NI}$$

(2) Selection subject to second circuit voltage.

The secondary circuit voltage must meet the requirements of standard instruments for protection and measurement. The secondary voltages are varied based on different wiring of voltage transformer, which can be selected according to table 4-2.

Table 4-2　Rated Voltage Selection of Voltage Transformer

Form	Primary voltage (V)	Secondary voltage (V)	Third winding voltage (V)	
Single phase	Connected onto the primary line voltage (e.g. Vv connection)	U_{Ns}	100	
	Connected onto the primary phase voltage	$U_{Ns}\sqrt{3}$	$100/\sqrt{3}$	Neutral-point indirect grounding system	$100/3$、$100/\sqrt{3}$
				Neutral-point direct grounding system	100
Three-phase	U_{Ns}	100	100/3	

Note: U_{Ns} indicates the rated voltage of system.

(3) Selection of type and model.

The type and model of voltage transformer shall be selected based on installation site and service conditions. For example, the oil immersed or cast type voltage transformer is generally used for the 6—35 kV PDUs indoors. The cascade electromagnetic voltage transformer is adopted generally for the 110—220 kV PDUs. When the capacity and accuracy class of the 200 kV and above PDUs meet the requirements, the capacitive voltage transformer is generally used.

(4) Selection subject to capacity and accuracy class.

The selected accuracy class of voltage transformer shall conform to the highest accuracy class of the supplied measuring instrument, and the wiring mode of voltage transformer shall be selected as per the wiring requirements of instrument and relay. Besides, the load shall be distributed at all phases to the greatest extent, and then the load size of each phase shall be calculated.

The rated secondary capacity (corresponding to the required accuracy class) S_{N2} of the transformer shall not be less than the second load S_2 of the transformer, i.e.:

$$S_{N2} \geqslant S_2$$
$$S_2 = \sqrt{(\sum S_{me} \cos\varphi)^2 + (\sum S_{me} \sin\varphi)^2} = \sqrt{(\sum P_{me})^2 + (\sum Q_{me})^2}$$

Where, S_{me}, P_{me} and Q_{me}—apparent power, active power and reactive power of each instrument;

$\cos\varphi$—power factor of each meter.

4. Selection of Buses, Cables and Insulators

1) Selection of Bus

(1) Bus material, type and layout.

The rigid bus is made of copper, aluminum and steel, and the bus is generally made of aluminum or aluminum alloy. Common flexible conductors include aluminium cable steel reinforced, composite conductor, bundled conductor and expanded diameter conductor, which are mostly used for outdoor PDUs.

The cross section of bus is rectangular, grooved or tubular. Rectangular conductors are often only used in the PDUs with the voltage of 35 kV and below and the current of 4,000A and below. Grooved conductors are often used for the 4,000—8,000A PDUs. Tubular conductors are used for the high current buses above 8,000A or for the PDUs of 110 kV and above.

Bus layout shall be determined based on the current carrying capacity, the level of short-circuit current, and the specific situation of PDU.

(2) Cross section selection of bus.

The cross section of bus can be selected based on the allowable long-term heating current or the economic current density. Except for the busbar of PDU, for the conductors with many annual load utilization hours, large transmission capacity, and the length over 20 m, the cross section is generally selected based on economic current density.

① Selection subject to allowable long-term heating current of conductor.

The formula is as follows:

$$KI_{a1} \geq I_{max}$$

Where, I_{max}——maximum continuous working current in the circuit in which the conductor is located;

I_{a1}——allowable current of conductor at the rated ambient temperature $\theta_0 = 25℃$;

K——comprehensive correction factor related to actual temperature and altitude.

② Selection subject to economic current density.

Cross section selection of conductor subject to economic current density will enable the minimization of annual cost. For different types of conductors and varied maximum load utilization hours T_{max}, the current density with the lowest annual cost exists, which is called economic current density J.

Economic cross section of conductor:

$$S = \frac{I_{max}}{f}$$

Where, I_{max}——maximum continuous working current during normal operation.

(3) Corona voltage validation.

Corona discharge will cause many adverse effects like corona loss, radio jamming, noise and metal corrosion. For the bare conductors of 110 kV and above, validation can be subject to no comprehensive corona on sunny days. The critical voltage U_{cr} of the bare conductor shall exceed the maximum working voltage U_{max}, i.e.:

$$U_{cr} > U_{max}$$

When the selected model of flexible conductor and the outer diameter of tubular conductor are more than or equal to the following values, corona validation is not required: 110 kV, LGJ-70/ϕ20; 220kY, LGJ-300/ϕ30。

(4) Thermal stability validation.

When verifying the thermal stability of conductor, if the influence of the skin effect coefficient K_s is considered, the minimum cross section of the conductor determined by thermal stability is as follows:

$$S_{min} = \sqrt{Q_k K_s / (A_k - A_i)} = \sqrt{Q_k K_s / C}$$

Where, C——thermal stability factor, $C = A_k - A_i$, the value of C is related to the material and working temperature of conductor. The selected cross section shall be more than or equal to S_{min}.

(5) Dynamic stability of rigid conductor.

Rigid conductors of various shapes are usually installed on post insulators, and the conductor may be bent or even damaged by the electrodynamic force from short-circuit impulse current. Therefore, the stress of rigid conductor shall be calculated based on the bending condition, while dynamic stability validation is unnecessary for flexible conductors.

If there are two conductors at each phase at least, when the short-circuit current passes through the conductor, the inter-phase stress σ_{ph} is generated on the cross section of conductor under the action of the inter-phase bending moment M_{ph}. In the meantime, the inter-bar stress σ_b is generated on the cross section under the action of the inter-bar bending moment M_b.

When the directions of σ_{ph} and σ_b are the same, the generated maximum stress σ_{max} is as follows:

$$\sigma_{max} = \frac{M_{ph}}{W_{ph}} + \frac{M_b}{W_b} = \sigma_{ph} + \sigma_b$$

Where, W_{ph} and W_b respectively indicate the inter-phase and inter-bar anti-bending section factors of conductor.

If σ_{max} is not more than the maximum allowable stress σ_{al} of the conductor, the conductor meets the dynamic stability requirements, i.e.:

$$\sigma_{max} \leqslant \sigma_{al}$$

(6) Conductor resonance validation.

Resonance validation is required for the conductors of important circuits (such as generators, transformers and busbar).

The following formula shall be used for calculation:

$$f_1 = \frac{N_f}{L^2}\sqrt{\frac{EI}{m}}$$

Where, f_1—inherent frequency of order, Hz;

L—span, m;

N_f—coefficient of frequency, N_f varies based on the number of continuous spans and supporting mode of conductor.

E—elasticity modulus of conductor, P_a;

I—secondary spacing of conductor cross section, m^4.

① Known span L of insulator.

When the first order inherent frequency f_1 of the rigid conductor is within the range of resonant frequency, the value of β is found out; When f_1 exceeds the range of resonant frequency, $\beta \approx 1$.

② Unknown span L of insulator.

When the actual span L of the selected insulator is not more than the maximum allowable span L_{max} of the insulator if the rigid conductor has no resonance, $\beta \approx 1$.

2) Selection of Cable

Power cables shall be selected and verified according to the following conditions:

① Cable core materials and model; ② Rated voltage; ③ Cross section selection; ④ Validation of allowable voltage drop; ⑤ Thermal stability validation. It is unnecessary to verify the dynamic stability of cables, which is guaranteed by the manufacturer.

(1) Selection of cable core materials and types.

Power cable cores are made of copper and aluminum, and the aluminum core cables are generally used in domestic projects. There are many types of cables, and cables shall be selected based on their purposes, laying methods and service conditions. Except for the single-phase cross-linked polyethylene cables or single-phase HV oil-filled cables used in case of 110 kV and above, the three-phase aluminum-core oil-immersed paper insulated cables, rubber insulated cables, polyvinyl chloride insulated cables, or cross-linked polyethylene cables are often used. The three-core and four-core (three-phase four-wire) power cables are often used. Heat-resistant cables are advisable in the places at high temperature. Flame-retardant cables should be selected for important DC circuits or emergency power supply cables. Band-armored cables are buried underground generally. Plastic sheathed cables shall be selected in the areas with moisture or corrosion. Cross-linked polyethylene cables shall be laid in the places with high altitude difference. With the development of material technology, more and more flame-retardant heat-resistant cross-linked polyethylene cables have been widely used. Oil-immersed paper insulated cables and oil-filled cables have tended to be obsolete.

(2) Voltage selection.

The rated voltage U_N of the cable shall be more than or equal to the rated voltage U_{Ns} of the located power grid, i.e.:

$$U_N \geqslant U_{Ns}$$

(3) Cross section selection.

The cross section of power cable is generally selected subject to the allowable long-term heating current. In case of the maximum load utilization hours of cable T_{max}>5,000 h and the length exceeding 20 m, selection shall be subject to the economic current density. The cross section selection methods of cables are basically the same as those of bare conductors.

In order not to damage the cable insulation and protective layer, a certain bending radius shall be maintained for the cable during laying. For example, the bending radius of the multi-core paper insulated lead-sheathed cable shall not be less than 15 times the outer diameter of the cable.

(4) Validation of allowable voltage drop.

The voltage loss $\Delta U\%$ shall be verified for the cable lines with long power supply distance and large capacity. It shall conform to $\Delta U\% < 5\%$ generally. For the three-phase AC, the formula is as follows:

$$\Delta U\% = 1731_{\max} L(r\cos\varphi + x\sin\varphi)/U$$

Where, U and L—working voltage (line voltage) and length of line;

$\cos\phi$—power factor;

r and x—resistance and reactance per unit length.

(5) Thermal stability validation.

Since the cable core is generally composed of multiple strands, when the cross section is less than 400 mm², $K \approx 1$, and the minimum cross section in line with the thermal stability requirements of cables can be simplified as:

$$S_{\min} = \sqrt{Q_k}/C$$

3) Selection of insulator

Post insulators shall be selected based on their rated voltages and types, and the short-circuit dynamic stability validation is required. Wall-through bushings shall be selected based on their rated voltages, rated currents and types, and the short-circuit thermal and dynamic stability validation is required.

(1) Selection of post insulator and wall-through bushing subject to rated voltage.

The rated voltages U_N of the post insulator and wall-through bushing shall be more than or equal to the rated voltage U_{Ns} of the located power grid, i.e.:

$$U_N > U_{Ns}$$

(2) Selection of wall-through bushing subject to rated current.

Rated current I_N of wall-through bushing shall be more than or equal to the maximum continuous working current in the circuit, i.e.:

$$I_N > KI_{\max}$$

Where, K—coefficient for temperature correction.

For the bus type wall-through bushing, due to no conductor, it is unnecessary to perform selection and verify the thermal stability based on this item. However, it is only required to guarantee the type of bushing and the size through the bus are fit for each other.

(3) Selection of types and models of post insulator and bushing.

The indoor, outdoor (or anti-fouling) product types in line with service requirements shall be selected based on the location and environment of the device.

(4) Thermal stability validation of wall-through bushing.

The heat effect $I_t^2 t$ of the bushing withstanding the short-circuit current shall be more than or equal to the heat effect Q_k generated by the short-circuit current through the bushing, i.e.:

$$I_t^2 t \geq Q_k$$

(5) Dynamic stability validation of post insulator and bushing.

Dynamic stability validation is required for post insulators and wall-through bushings. In case of short circuit, if the force on the post insulator (or wall-through bushing) is equivalent to the average of the electrodynamic force F_{max} on its adjacent transconductors, the flexural fracture strength F_{de} of the post insulator (or wall-through bushing) shall meet the following requirements:

$$0.6F_{de} \geqslant \frac{H_1}{H}F_{max}$$

参考文献

[1] 高建. 电气设备检修[M]. 成都：成都时代出版社，2019.

[2] 杨迪. 变电检修技能培训教材[M]. 北京：中国电力出版社，2019.

[3] 姜聿涵. 变压器检修技能培训教材[M]. 北京：中国电力出版社，2019.

[4] 姜聿涵，杨冰."一带一路"变电设备检修专业培训教材 变压器检修（结构及附件篇）（中英文对照）[M]. 北京：中国电力出版社，2021.

[5] 华章，雷春."一带一路"变电设备检修专业培训教材 变压器检修（电气试验篇）（中英文对照）[M]. 北京：中国电力出版社，2021.

[6] 邓常飞，祝捷."一带一路"变电设备检修专业培训教材 变电检修（故障处理篇）（中英文对照）[M]. 北京：中国电力出版社，2021.